Extrater

Telepathy

Earthside & Spaceside Visits to the Moon

UltraSuperSecret UFO Activities

Coverups

Ingo Swann reveals a long-held secret series of experiences with a "deep black" agency whose apparent charter was simple: UFOS and extraterrestrials on the Moon and worries about ET telepathic/mind control powers.

All the while, he explores the fact that we officially know far more than we're admitting about the Moon…and Mars.

Ingo Swann (1933-2013) was an American artist and exceptionally successful subject in parapsychology experiments. As a child he spontaneously had numerous paranormal experiences, mostly of the OBE type, the future study of which became a major passion as he matured. In 1970, he began acting as a parapsychology test subject in tightly controlled laboratory settings with numerous scientific researchers. Because of the success of most of these thousands of test trials, major media worldwide often referred, to him as "the scientific psychic." His subsequent research on behalf of American intelligence interests, including that of the CIA, won him top PSI-spy status.

His involvement in government research projects required the discovery of innovative approaches toward the actual realizing of subtle human energies. He viewed PSI powers as only parts of the larger spectrum of human sensing systems and was internationally known as an advocate and researcher of the exceptional powers of the human mind.

To learn more about Ingo, his work, art, and other books, please visit: **www.ingoswann.com**.

PENETRATION:

SPECIAL EDITION

UPDATED

THE QUESTION OF EXTRATERRESTRIAL AND HUMAN TELEPATHY

A BIOMIND SUPERPOWERS BOOK
PUBLISHED BY

Swann-Ryder Productions, LLC
www.ingoswann.com

First edition BioMind Superpowers Books.

Cover art by: *Connecting Up* by Ingo Swann © Swann-Ryder Productions, LLC.

Internal art from Shutterstock.com: Moon Surface © HelenField | Spacecraft Metal Gate © agsandrew | Lighthouse on the Moon © muratart | Watching You © agsandrew | The pyramids and face on Mars © HelenField

ISBN-13: 978-1-949214-64-2

PENETRATION:

SPECIAL EDITION

INGO SWANN

PUBLISHER'S NOTE

Inserted within the draft materials for *Penetration* was a chapter Ingo wrote and numbered "9." This missing chapter introduces us, the reader, to two psychic probes of Mars: the first of these Ingo undertook with Harold Sherman in 1975 and was witnessed by Dr. Janet Lee Mitchell and Dr. Stanley Krippner and his wife Lelie; and the second in 1984, for which Ingo organized a group effort.

From reading "9," it is clear that Ingo intended to use this chapter as a launching pad for an entire section on Mars. However, and for reasons we will never know, he ultimately decided against this path. Instead he self-published *Penetration* in 1998 with a focus exclusively on the Moon.

Ingo built the final version of *Penetration* around three sections: the first is the retailing of his adventures with a secret defense/intelligence-type organization; the second is his treatise on the tangled confusion around the Moon; and the third, contains his discourse on human and extraterrestrial telepathy. But at the heart of *Penetration* is, as Ingo writes, the fact that telepathy is an element of what he calls "consciousness universal."

In "9," Ingo asserts that there is a bigger question at play—the question as to "why do mass-consciousness humans, as it were, mass-consciously almost 'conspire' to avoid certain issues, and consistently so?" This inquiry, deep within Ingo's own awareness, is one that he would ponder until his passing in 2013.

Evident in "9" is Ingo's desire to share this interrogation back in 1998. And so that is what we did in 2019. However, subsequently, we uncovered Ingo's missing 1984 Mars session notes within Harold Sherman's Archives at the University of Central Arkansas. So, while Neptune goes direct in Pisces, removing the veils, and providing an awakening of sorts, and during the full moon and lunar eclipse in Gemini, known for shining light on things ignored in the past, we have decided to bolster Ingo's question by including these notes, in their original form, along with "9," the Introductions by Dr. Krippner and Dr. Mitchell, and an Afterword by Thomas M. McNear, Lieutenant Colonel, U.S. Army (Ret.) in this updated Special Edition, all within an aptly named section entitled **Subscript**.

TABLE OF CONTENTS

PENETRATION
(definitions of)

1) To pass into or through;
2) To enter by overcoming resistance;
3) To pierce;
4) To see into or through;
5) To discover the inner meaning or contents of;
6) To pierce something with the eye or mind;
7) Having the power of entering, piercing or pervading;
8) The act of entering so that actual establishment of influence is accomplished.

PREFACE

THIS BOOK IS DEDICATED TO "MR. AXELROD," WHEREVER AND WHOEVER HE IS

This small book is divided into three parts. And it is with regard to the first part that I join a very long list of those who have seen and experienced things they cannot prove happened. The second part is on much firmer ground. It is largely a brief synopsis of spectacular data and facts about the Moon that have achieved exposure elsewhere, and which provide evidence that the Moon is a very interesting place, indeed.

I have selected only small portions from all of the unusual lunar information available but have provided sources in the bibliography for those interested in more copious details.

The third part begins with a presentation of certain social phenomena regarding the problems of telepathy that can be factually identified. These, however, set the background for the strange and surprising scenario they lead into, although the scenario is admittedly speculative in nature.

Some have advised me not to publish this book on the grounds that it challenges those echelons of conventional credibility that lasciviously get off on deconstructing those unfortunates who experience what they can't prove.

I have been mindful of this opportunistic factor for several decades. But my age is advancing, and I have become interested in recording and wrapping up my active research into Psi phenomena in order to pursue less stressful vistas.

As I have discussed in other writings, I've always been interested in Psi phenomena, and beginning in 1970 it chanced that opportunities to extend that interest in depth were made available.

Anyone with more than a mere superficial interest in Psi phenomena must of course encounter the rather smelly morass of social resistance whereby the authenticity of those phenomena is methodically deconstructed, thus suspending them in doubt.

This social resistance, even if smelly, has largely been successful in destroying all concerted approaches to Psi phenomena. This success is specifically active within high strata of societal power, and which strata are otherwise entirely disinterested in what lesser mortals DO experience along these lines.

Why it is that governing societal factors need to deconstruct the provable existence of at least some vital Psi phenomena is therefore something that needs to be examined and understood.

Along these lines of inquiry, the existence and methods of the machinations against Psi *development* can easily be brought to light. But the *reasons* that govern the implementation of the machinations none the less remain obscured.

Thus, the societal resistance to Psi breaks neatly into two aspects: to prevent Psi development; and to keep obscure the actual reasons for doing so.

One reason for the blanket suppression which has been offered up by many before me is that effective formats of Psi would disturb any number of social institutions. Those institutions would feel "threatened" by developed formats of, say, telepathy, which might thereafter be utilized to penetrate their secrets.

There is some rather clear truth in this. Indeed, it is because of this truth that some echelons of humans are at war with the Psi potentials of the human species because those echelons have motivations they would prefer never to be disclosed via Psi penetration.

If this is the case, the chief preventive measure would be to stamp out altogether any real understanding of Psi. Indeed, something like this has taken place.

And the cognizance of the nature of the situation might remain more or less being defined as humans in conflict with their own Psi potentials because Psi penetrates secrets.

Indeed, on my part for a long time I assumed that this was the beginning and end of the story regarding the methodical suppression of Psi by high societal echelons such as represented by government, science, academe, and media.

As it happened, however, the events described in Part One of this book occurred beginning in 1975.

These are the events I can't prove. None the less they made somewhat visible another possible aspect that might be factored into the odiferous suppression of Psi that was already familiar to me.

This aspect required that I introduce two unusual terms: EARTHSIDE and SPACESIDE. These refer, of course, to Earthside intelligence and Spaceside intelligence.

The central hypothesis of this book is that if developed Psi potentials would be an invasive threat to Earthside intelligences, then developed Earthside Psi would also be a threat to Spaceside intelligences.

After all, in that telepathy, for example, is invasively defined as reading minds, the distinction between reading Earthside minds and Spaceside minds would be very narrow.

The only real problem in considering this is whether or not Spacesiders exist.

I have decided not to enter into the relevant debate about this issue but direct the reader to the copious literature already existing, with special regard to the weekly UFO ROUND-UP that can be located in the Internet (see bibliography).

The inclusion in this book of the story I can't prove is not being offered as evidence about the existence of Spaceside intelligence, but because the reader deserves to know why I have concluded there is far more to telepathy than commonly conceived in Earthside terms.

In this, the thinking proceeds from actual experience and not from analyzing the information packages presented in the works of others.

The works of others, of course, have proved to be valuable in the long run, and they certainly introduce a modicum of authenticity that would otherwise go completely missing.

In the end, though, the authenticity of my personal, unprovable, experience probably doesn't need to be considered all that much because the drift of accumulating information is inexorably leading to establishing the authentic existence of extraterrestrial intelligences anyway.

One factor that won't be apparent throughout this book is the large amount of time (years actually) it took to achieve the synthesis of the factors presented. I tend to be a rather slow thinker and am sometimes even slower on the up-take.

I had originally intended to include a lengthy discussion regarding the probability that telepathy might be a universal "language" system of some kind that operates through consciousness entities everywhere.

I briefly allude to this in Part Three, but otherwise have decided to include that discussion in another work because it needs a larger information basis that includes the nature of energy organisms.

But I feel obliged to comment on some of the reasons I decided to go ahead with the book after so many years have passed.

In late 1990, I read a well-documented report of a large UFO craft sighted in the former USSR.

The report indicated that the sighting was attested to by General Igor Maltsev, chief of the main staff of Air Defense Forces, and published in *Rabochaya Tribuna, 19,* dated April 1990.

The report quoted General Maltsev as saying: "I am not a specialist on UFOs and therefore I can only correlate the data and express my own supposition. According to the evidence of these eyewitnesses, the UFO is a disk with a diameter from 100 to 200 meters. Two pulsating lights were positioned on its sides..."

The article went on to state that UFOs are piloted craft and contradicted the suggestion that they are mere atmospheric phenomena. If the sighted craft was indeed 200 meters, it was about 650 feet, or somewhat larger than a football field.

Meanwhile, there were other notable sightings elsewhere, and video footage was being obtained regarding a lot of them. Such reports got me ruminating about my 1975 experiences, with the result that I decided to write them down before my memory began deteriorating more than it already had.

Between 1976 and 1990, I gradually concluded that Earthsiders and Spacesiders didn't seem to have much in common with the exception of telepathy.

By all contactee and abductee accounts, telepathic capacities seem to be well-developed by the ET's, but remained quite undeveloped Earthside.

I expanded the narration of the events to include some fundamental considerations of telepathy, and which theorized WHY development of telepathy is suppressed Earthside.

In due course, I showed the manuscript to my then literary rep, who got excited about it, and thought that its successful publication was a sure and easy thing.

Over twenty publishers turned it down even in the face of the fact that much UFO-ET stuff ranging from bullshit to the sublime fantastic was otherwise being published everywhere.

This blanket rejection on such a large scale remains, as it were, mysterious. Perhaps it can be interpreted as some kind of subtle, large-scale media control. But one possible explanation might be that as outrageous as the tale and telepathic considerations are, something in them moves too close to Someone's comfort.

In any event, because of frustration, embarrassment, etc., I abandoned this book project. And some more years passed. In about March, 1998, however, certain articles and TV reports centering on ET possibilities began circulating, among which were a few entitled "Astonishing Intelligent Artifacts (?) Found On Mysterious Far Side Of The Moon."

Then, via a report on the Internet authored by David Derbyshire, dated May 14, 1998, it seems that a "24,000 mph UFO" buzzed

Britain on May 13,1998.

THIS craft was tracked by the Royal Air Force and the Dutch Air Force. It was "triangular" and "as big as a battleship. About 900 feet long." British and Dutch interceptors were sent aloft. The Big Thing left them in the mists and went who knows where?

Thus, there are recent authentic reports of UFOs, and indeed they seem present Everywhere, and even boldly reveal themselves to the lenses of Camcorders world-wide. That the UFOs are driven or managed by Spaceside intelligences simply must be taken for granted. And if they have achieved high technological control of consciousness that is commensurate with the high technology of their craft, then I'll bet they are very good at what we Earthsiders refer to as telepathy.

ULTRA-SECRET
GOINGS ON

INVOLVEMENT IN PSI RESEARCH

The sequence of strange events narrated in this book took place because of my involvement with Psi research, which began out of the blue in 1971 when I was thirty-seven years along.

My life might have flowed along lines presumably more gratifying in mundane but more comfortable ways had I never volunteered to be an experimental subject in Psi research labs.

In these experiments, there were high and low points, successes, and failures. And there was the opportunity to meet with many fabulous and wonderful people.

But when one enters into Psi research, one also enters into a narrow cultural subset rather steamed up with high stress factors, intrigues, mainstream confusions, fear and apprehension, internecine warfare, and largish clumps of idiocy.

Additionally, Psi research subjects (guinea pigs) are nonentities who are expected to exhibit Psi manifestations. At the same time, they are supposed to know nothing, think nothing, suppose nothing—because the job of knowing, thinking, supposing belongs to the researchers.

The subject is something like a computer chip being tested to see if it can perform in the ways wanted. If the chip doesn't perform in the ways wanted, then it is tossed aside into the big pile of anonymous chips that have likewise failed.

It is thus that the laboratory lifespan of a test subject usually does not exceed three months, and during that time they have to undergo endless repetitive testing. One of the major outcomes of this is usually bottomless boredom.

The appearance of boredom is deadly in Psi research—because a bored chip gets into a state of apathy or non-interest, after which its delicate circuitry fizzles.

I knew most of this in advance, largely because Psi phenomena had always been of endless interest to me, and I had done a great deal of organized reading and study. So I had no expectations at all that my allotted three months would somehow turn into nineteen years.

The major reason had not much to do with the Stygian climes of parapsychology itself. Unknown to almost everyone at the time, the American intelligence services became worried about, of all things, possible development of "psychic warfare advances" in the (now former) Soviet Union.

The intelligence services are heavy players, and because of the Soviet Psi threat they more or less required an active picture of Psi potentials somewhat larger than standard parapsychology could provide. Because of these unusual circumstances, I got dragged into several years of work in this regard.

But this meant that I also got dragged into realms of often idiotic secrecy, into endless security checks conducive of paranoia, into all kinds of science fiction dreamworks, into intelligence intrigues whose various formats were sometimes like toilet drains, and into quite nervous military and political ramifications.

My participation in this long-term affair had its ups and downs—and was to involve hundreds of complicated situations, circumstances, and events of various kinds—of which those narrated in this book were only one kind, albeit the most stressful and mind-boggling.

To get into the elements of the narration, it is necessary to briefly outline what led up to them.

In late 1972 the Central Intelligence Agency funded a small, tentative research project at Stanford Research Institute. The project at SRI was headed by the physicist, Dr. H. E. Puthoff, and I was invited to travel to California to participate in it.

The purpose of the small project was to discover one ESP phenomenon that was capable of being reproduced at will. This was the kind of experiment notoriously missing in parapsychology, but in which I had been somewhat successful earlier.

The project was given eight months to produce something along these lines. So, thereupon began yet another daily exercise involving hundreds of experimental trials. These proceeded up and down in terms of what was being tested, but ultimately DOWN into boredom so cloying that it was hard to face yet another day of it.

In early April, 1973, in an effort to emerge from the daily boredom of repetitive testing (which induces a flatline of ESP activity), I suggested that we once in a while do something far out, something that might reintroduce a sense of adventure, excitement, and enjoyment.

The planet Jupiter was literally far out. NASA had earlier launched Pioneer 10 and 11 to fly-by that planet, and information telemetered back by the two crafts would undergo technical analyses. Information from Pioneer 10 would commence in September, 1973.

The only real difference between Jupiter as a "target," and mundane target objects in the next room, was its distance from Earth. But for me there was another difference. It would be exciting to try to extend one's ESP to the planet, a form of remote viewing. Jupiter was more remote than the next room—and there might be a thrill of "traveling" in interplanetary space.

But there was yet another difference. Those locked into conventional research mindsets are usually nervous about novel experiments. Conventional mindsets tend to take themselves somewhat seriously, so there is usually resistance to non-conventional experiments.

The resistance is usually first manifested by tar and feathering the proposed experiment (and everyone involved) BEFORE it takes place. If that doesn't squelch the experiment, then it is merely declared ridiculous and laughed out of Sciencetown.

Is not a psychic mind trip to Jupiter laughable?

My colleagues at SRI were, to put it mildly, not interested in being laughed out of town. But I had become quite gloomy since failure-by-boredom was just ahead.

So I had a choice of (1) being laughed out of town, or (2) boredom which clearly could flat-line ESP faculties. The resistance to the Jupiter "probe" was overcome when I said, "I quit, and you can return what's left of the money to the funding clients."

In any event, I felt it would be interesting to see if the remote viewing data acquired in April, 1973, might somewhat match the data later revealed by NASA's craft beginning in September, 1973.

The thrill of the idea was to get psychically to Jupiter before the NASA vehicles did. If this worked even somewhat, it was a kind of psychic one-upsmanship. The experiment was done on personal time, on a Saturday, a non-working day.

But it was wrapped in very stringent protocols. At first, the very-long-distance (VLD) experiment was not to be an official one. But the remote-viewing raw data had to be recorded somehow, so that it could be established that it existed prior to the NASA vehicles getting to the planet.

So, at the conclusion of the experiment, copies of the raw data were circulated far and wide, offered to and accepted by many respected scientists in the Silicon Valley area, including two at Jet Propulsion Laboratories. Some scientists, of course, thought the entire idea ridiculous, but these were fewer than one might expect.

For the experiment to be considered successful in any way, the remote viewing data had to include impressions of factors that were not known about the great planet—lest one be accused of reading up beforehand.

As to the raw data itself, this ended up consisting of one page of sketches, and two and a half pages of verbal observations.

The raw data yielded thirteen factors, and only thirteen, all of which were scientifically unanticipated before they were confirmed by later analysis of the scientific data. These raw data factors are enumerated below, accompanied by the dates they were confirmed.

- ∅ The existence of a hydrogen mantle: Confirmed September 1973, again in 1975.

- ∅ Storms, wind: Confirmed 1976 as to dimensions and unexpected intensities.

- ∅ Something like a tornado: Confirmed 1976 as strong rotating cyclones.

- ∅ High infrared reading: Confirmed 1974.

- ∅ Temperature inversion: Confirmed 1975.

- ∅ Cloud color and configuration: Confirmed 1979.

Ø Dominant orange color: Confirmed 1979.

Ø Water/ice crystals in atmosphere: Confirmed 1975.

Ø Crystal bands reflect radio probes: Confirmed 1975.

Ø Magnetic and electromagnetic auroras ("rainbows"): Confirmed 1975.

A planetary RING inside the atmosphere: Confirmed 1979, not only as to its existence, but as being inside the crystallized atmospheric layers.

Liquid composition: Confirmed 1973, 1976, as hydrogen in liquid form.

Mountains and solid core: Still questionable but suspected as of 1991.

Six of these thirteen factors were given scientific substantiation by 1975, which is the year that the events narrated in this book begin.

It needs to be pointed up that before it was actually discovered in 1979, most scientists flatly denounced the possibility of the RING, but which had been sketched in the raw data acquired in 1973. And just recently the existence of more refined rings has been confirmed.

For me, the Jupiter experiment effected a cure of my experimental doldrums for a number of reasons.

For one thing, the trip and the sightseeing there were awesome experiences. This was a kind of profound aesthetic impact that can inspire one for many of years.

For another thing, as confirmation feedback began coming in during September, 1973, in the form of scientific announcements, the gossip lines shifted from cold ridicule and began heating up. Lots of notables began coming to lunch at SRI in order to get grounded with the possibilities.

For yet another thing, the CIA, of course, was interested in the possibilities of psychic spying. Although the planetary experiment had not been done on the Company's funded time, it now seemed that the project at SRI was excitedly on the right track.

The Jupiter Probe also received wide media coverage, although not in scientific journals, of course. But then there are all types of people who view science much in the same way that science has traditionally viewed parapsychology.

I now wish to mention an aspect that might go missing otherwise, and does go missing, rather conveniently, as regards a lot of psychic claims and posturing. This has to do with the matter of what are referred to as positive feedback loops.

It is not hard to comprehend what these consist of. One word will do: Confirmation—in some or any form.

A "psychic" says thus and so, after which one needs to look around for some kind of hard evidence that supports the real-time facts of what has been said.

As far out as the Jupiter experiment was, it was based and designed AGAINST expected feedback loops.

The feedback was in the form of the information telemetered back to Earth by the NASA vehicles flying by the planet.

As it turned out, among those taking an active interest in the possibility of interplanetary spying was a group so clandestine that it could be characterized not merely as a deepest black project, but as an entirely invisible one.

It was this group, or whatever it might be called, that I met up with in the early part of 1975.

ENCOUNTERING THE SPOOKIEST OF SPOOKS

A bout two years after the Jupiter probe I received a telephone call during late February, 1975, from a certain highly-placed functionary in Washington, D.C.

I had met him on social occasions, and we had rather enjoyable conversations since he had a deep interest in Psi research.

I both admired and respected him. He was forthright about his unusual interests, and he dared to swim against the surface currents of that mighty river called "prevailing opinion" which could damage even very high reputations in the Washington maze.

But in his telephone call to me, my friend was somewhat less than forthright, as the following conversation drawn from memory indicates.

"A Mr. Axelrod is going to telephone you," he said. "If you can manage to do so, would you try to do whatever he asks, and ask no questions yourself."

After a pause, I asked: "Well, who is Mr. Axelrod?"

Now there was a pause at his end of the telephone. Then: "I can't tell you because I don't know myself. But it's important, VERY important, very URGENT that you agree to do what he asks.

"I can tell you nothing more, so please do NOT ask. Just do what he wants. And, whether you do or do not, we will never refer to this conversation again. I must ask you in friendship never to refer to me about this in any way." After which, my friend expressed a quick passing interest in how I was doing, and then virtually hung up on me.

Although my contact was usually jolly, he had seemed, well, a little uptight. But otherwise, this type of thing was not entirely unusual in my new career of Psi research.

Many had approached me, some of whom asked for anonymity,

some using fake names—such as police emissaries and detectives who wanted inputs regarding difficult crimes, a few scientists with research stoppages, an art director of a famous museum which had misplaced a valuable painting.

Desperate people do desperate things—such as consulting psychics—even some Presidents whose interactions with seers are documented.

In this somewhat less than open manner began a chain of mind-boggling affairs which excited me on the one hand, yet ultimately made me QUIVER as if I suddenly found myself standing between two realities neither of which seem quite real.

As it turned out, in spite of the alleged urgency, the mysterious Mr. Axelrod did not telephone until about four weeks later. And when he did, it was about three in the morning. The call jolted me out of a sound sleep, so, of course, at first, I didn't quite remember who he was.

After we got that straightened out, he asked: "Can you get to Washington by noon today? I realize this is short notice, but we would be very appreciative if you can. We will reimburse you for your time and all your expenses."

I was just about to ask why I should get to Washington by noon, when I remembered that my friend had been very insistent that I not ask questions. So, I said I would take the air-shuttle or something. "Good," Mr. Axelrod said, "but we cannot meet you at the airport. Are you familiar with the Museum of Natural History at the Smithsonian?"

I said I was. "Good," he replied. "As soon as you arrive, go there and stand near the elephant in the central rotunda. Be there at noon. You will be contacted. Just do exactly as your contact asks. My only requirement is that you tell no one where you are going. If you feel you cannot do that, please say so now and we will forget about this."

I sat in silence. "Is that OK with you?" he asked. "Yes, I suppose so." But I couldn't resist one question, which seemed a logical one. "How will I recognize who is supposed to contact me?"

"Don't worry. We know what YOU look like." And Mr. Axelrod then hung up without so much as a good-bye.

I got out of bed, made some coffee, chain-smoked some cigars, and sat contemplating the noisy darkness outside my windows (New York City is always noisy.)

I was beginning not to like this at all and were it not for my highly-placed acquaintance in Washington, whom I respected, I am quite sure I would have decided the whole affair had suddenly become too questionable to proceed with.

The world, back in 1975, it should be remembered, was in the grips of the Cold War. My research colleagues at Stanford Research Institute and I had speculated that the Soviet KGB would naturally be interested in what we were doing. And in our more dramatic considerations of this possibility, it was even speculated that one of us might get kidnapped or worse by that infamous, but very smart organization.

Well, I decided, if I got to Washington early enough, I could once more view the magnificent collection of minerals and crystals housed in the Museum of Natural History. Doing so had turned me on for years.

So, as the sun was rising in the rather cool late winter weather, I made my way to LaGuardia Airport, and got aboard the next air-shuttle to Washington, about a fifty-minute ride.

I arrived with plenty of time to spare. In fact, the museum wasn't yet open, so I got some coffee and a roll from one of the vendors in the Mall and smoked some more cigars.

Needless to say, even when viewing the three-foot crystals, and looking at the egg-sized precious gems, my mind was not at all on them. In fact, I was sweating most of the time. Nerves? Apprehension?

Finally, I went to the mezzanine floor which encircles the museum's great rotunda. As covertly as possible (to my way of thinking, anyway) I surveyed the floor below-in the center of which stands the famous stuffed elephant with his absolute majesty entirely visible to all who take a moment to consider it.

Assuming I was supposed to act like any other tourist, at noon I was standing in front of the giant elephant, pretending to have interest in it.

Behind me a voice spoke. "Mr. Swann?" I turned and was

immediately handed a card which read: Please do not speak or ask any questions. This is for our safety as well as yours.

If I had not been convinced before that I was getting into something suspicious, I was now certain. For the guy who handed me the card stared at me with burning green eyes, which clearly indicated he meant business. I didn't dare speak.

He was young and looked like he could have posed for a Marine recruitment poster—which is to say, he was masculine and military. I sensed he was serious, and that his complacency concealed an ability to kill very quietly.

But even more astonishing was the fact that there were TWO of them which as far as I could tell were twins.

Dozens of museum visitors were flowing all around us.

After reading the card, I blinked. The first guy pulled out a photograph which I could see was of me. He studiously compared the photograph face to mine.

He then took my hand as if he was shaking it and compared the tattoo on it to another photograph—the tattoo I had gotten as a result of a rather drunken desire in 1962.

He then nodded to his duplicate, who had been watching the rotunda in a professional way, and this twin came over and repeated the sequence.

They then both initialed what seemed to be a kind of check-sheet in a small address booklet.

All this took place in a few moments, and none of the spectators passing to-and-fro around the elephant seemed to notice.

The first twin then nodded and indicated the main door to the museum. I followed, with the other twin just behind me.

We marched directly to the curb and got into a rather non-descript car boldly waiting in a No Standing zone.

The driver was a woman who, it seemed, deliberately never looked at me.

The car was large and blue, a little dirty on the outside, but immaculate inside. The twins sat in the back on either side of me. One of them pulled out another card: Please do not speak. You may smoke if you want. Which I gratefully did. I was wet in my armpits.

As it turned out, we were, I think, the center car between two others that stayed with us as we moved through the clogged streets.

Once we left Washington proper, the twin pulled out yet another card which read something like: Please do not take it personally, but we are required to check your person and clothing for weapons or bugs.

What could I do? They then proceeded to check everything, even unzipping my pants, and peeking briefly into my shorts. After which, they both initialed the mysterious check-sheet.

I had never been treated this way before. I had never been frisked. I wanted to be outraged. However, I didn't dare move or open my mouth, except to puff on a cigar.

By this time, I hardly knew where we were. We seemed to be heading in what I thought was the direction of the CIA headquarters hidden behind trees. I thought that might be our destination, but we zoomed past it, picking up speed.

Then came another card: You are being taken to a heliport for further transport. Before we reach it, we will place a hood over your head. It will be removed at your destination. If you are hungry, sandwiches are available.

At this point, I thought of doom. Yet, silly as it seemed, I WAS hungry, even if my stomach was in knots. I ate. Although my hands were now shaking a little, the twins pretended not to notice.

Well, I concluded, I'm either really being kidnapped, or whatever is about to go down IS something extraordinary.

About twenty minutes later, the first twin produced the promised hood, and I spent the next part of my voyage in stuffy darkness.

Not long after, the car ground to a halt. I was helped out, and with the two twins firmly grasping my arms was soon seated in a chopper. We took off almost before we were strapped in.

This part of the voyage seemed to last about a half hour, but I can't be sure. Shortly we bumped to a sudden landing. I was helped out and walked what seemed a longish distance. I then heard a door wheeze closed, and we DESCENDED. So, I knew we were in an elevator.

I was now turned around in the elevator, and when it came to a halt, after what I took to be quite some distance down, a door apparently opened up and we walked straight out.

Next, I was physically turned around several times by the twins. After about ten turns, we marched along, at one point seeming to go down a ramp. Shortly I was pushed down onto a chair.

At this point a voice actually SPOKE the first words I'd heard in this whole threatening goings-on.

"I'm going to remove your hood now, Mr. Swann, and thank you for coming as well as putting up with our procedures."

I was, well, terrified by now, and don't mind admitting it.

DERGROUND, OMEWHERE

With the hood removed, eyes watering a little, I found myself in a dimly lit room. The twins were nowhere to be seen.

The VOICE said: "I am Axelrod, which is not my real name, of course, as you must have figured out."

Mr. Axelrod was a jolly-guy type, smiling, with kindly eyes, but dressed in a dark-green jumpsuit of some kind. He reminded me of a certain Captain McBee I had worked with while in Korea.

He continued: "I can answer no questions as to where you are or what we represent, but beyond that I am at your complete disposal with regard to anything that pertains to the task ahead."

Mustering what dignity I could, which really was not much, I croaked out: "Well, what task, then?"

Mr. Axelrod smiled. "First, there are some procedural matters. We will reimburse you for your expenses and provide what we will call an honorarium. Would a thousand dollars a day be suitable? We will provide this in cash before you leave." "A DAY!" I croaked again. "How many DAYS?"

"Well, we have heard you work best in the morning, and as it is now in the afternoon, we will begin the task tomorrow morning at any time which suits you. After that, we will wing it a little."

A thousand bucks a DAY!!! I perked up—and stopped croaking —and even tried to say something sensible. "So, if you know about the morning thing, then you must be very familiar with our procedures out at Stanford Research Institute."

"We know a great deal about you, Mr. Swann. You seem to be an exceptional man, and of course it is your psychic gifts we want to try to employ with regard to the task."

"My 'psychic gifts', as you must then know, are very undependable. I work only in experimental situations, and I hardly think anyone should risk anything really serious on them."

"We understand fully, Mr. Swann. We do not see the task as a risk, so do not feel stressed about that.

"The second preliminary. We would like to ask you never to reveal any of the details about any of this, including your presence here. If the circumstances were otherwise, we would ask you to sign a secrecy agreement. But, bluntly speaking, we exist without leaving a paper trail regarding our mission."

Mr. Axelrod paused to let that sink in, and then continued. "Yet, without such an official secrecy oath, you will not be legally bound to secrecy. What we would hope, then, is that you will agree not to reveal this sequence for at least ten years hence.

"I can assure you there are very good reasons for this, but after ten years our mission will have 'disappeared', as it were.

"If you cannot see your way clear to making and upholding this agreement, we will give you a good dinner, discuss remote viewing, and get you back to New York by late tonight."

For the record here, other groups had invited me to work on many other kinds of sensitive projects and had even signed non-disclosure agreements. So, except for the ultra-secrecy of this one, which I thought merely overly dramatic, it was not all that unusual.

Although I was hot on the trail of $ 1,000-days, I frowned at Mr. Axelrod. "I guess you knew I would accept, or I would not be here now, would I?"

"Good, very good, then. We have specific procedures here. We will work in this room, if that is suitable. There is an adjoining room with a bed, and it's comfortable. It has a TV you can watch.

"You will see only myself, and the two who brought you here. They will be your constant companions when you are not with me. One will spend the nights in this room, and the other will be stationed directly outside the door. They do not know why you are here, and they need not know.

"If you need exercise, we have a small gym. We have shorts and gear and a small pool if you want to swim. If you have any kind of special food preferences we believe we can supply them. Just ask for what you want. You smoke Tiparillo cigars. We have some for you, as well as better ones if you wish. Can you work under these circumstances?"

I hardly knew what to say by this time.

So, bravely, I ventured, "I guess that depends on the work...or the task, or whatever it is." Then: "I know I am not supposed to ask anything, but are those two guys really twins?"

Mr. Axelrod smiled again. "What do you think?"

"I think they are."

"Well, then that is resolved, isn't it? Did you enjoy the geological specimens at the museum this morning?"

I decided to ask no more questions. Presumably, I had been observed ever since I left New York. Whatever was happening must be important since it obviously was costing someone a great amount of dollar/man-hours.

"Well, then, may I call you Ingo? And you must call me Axel. Tell me about remote viewing."

I decided to relax. "Well, as you may know, I did my first extensive clairvoyant experiments at the American Society for Psychical Research in New York with a woman named Janet Mitchell and, of course, with Dr. Karlis Osis, the director of research there.

"After a while, I got bored with trying to see targets in boxes and the next room. One day I decided to see what else I could see, and found I thought I could see people going down the street outside.

"One day, I thought I saw a woman dressed in orange and green walking along the street. We rushed downstairs just in time to see orange and green disappearing around the corner.

"There was really no way I could have seen her visually since I had been sitting in a closed room. I got to thinking. I proposed then a larger experiment.

"I would try to see things at greater distance, provided we could figure out a way to get easy feedback about what was being seen.

"We thought about this for a while, and finally decided that I could try to see the weather going on in major cities, and then call up their weather numbers to see if I was correct or not."

"How did you specify which city?" Axel asked.

"Well, we decided that Janet would compile a list of cities and select one at random. She would then say: the city is such and so, go there, Ingo, and see what the weather is doing.

"After I had said what the weather was, Janet would pick up a telephone, call long-distance and get the current weather report.

"This didn't work too well at first, but I suggested we try a number of times. Finally, we got a number of hits in a row.

"For example, she gave me Phoenix as a target. I saw it was raining there, or at least had just rained. Sure enough, Phoenix had just had a rainstorm, which was unusual because they don't often have any. Well anyway, we did this for a few days with pretty good results.

"Since these cities were remote from New York, we decided to refer to this kind of experiment as remote viewing. This started in December 1971. That's how it all got started."

Axel had his fingers pressed against his lips. He was no longer smiling and seemed pensive. So, I asked: "I gather you want me to try to remote view something?"

"Oh, absolutely, absolutely," he responded, resuming his smile. "After the American Society, then, you went to SRI and developed a coordinate-ordinate system for remote viewing?"

"Well, that came about because we wanted to try to view sites around the world. The CIA was interested, you know.

"When we tried to target the cities by their names, we realized that the name had too many clues which might aid me in identifying the target.

"We felt that skeptics and critics would point this out, making our work useless. So we felt we couldn't do that kind of experiment. After all, if you say 'New York' for example, anyone would know enough to say they see skyscrapers, and so forth.

"But one day, in 1973, I was swimming in a pool in the apartment complex where I was staying in Mountain View, which is near Palo Alto and Menlo park where SRI is located. I had been wondering how we could identify a distant target in some other fashion than by its name.

"In the water, I lay back against the edge of the pool and tried to envision something which had escaped me. I suddenly saw a map with coordinate-ordinates on it, you know this degree East and that degree North. A 'voice' of some kind said (in my mind, of course), 'try coordinates'.

"So, I got the idea that if someone gave me a set of coordinate-ordinates they might act as a focus of some kind. At first my SRI colleagues thought this was silly, but I insisted we give it a good try. At first this didn't work too well either, but after about fifty tries, it began to pay off."

"Can you explain why coordinates seem to work better than other ways of specifying a target?" Axel asked.

"No one understands this at all, and neither do I. The criticism is that coordinates are only arbitrary sets of numbers and as such bear little real meaning to the actual physicality of the site.

"But my explanation, if it is one, is that people do find their way around the world by using coordinates. And since this is so, then there is no real reason why one cannot use them to find their way in a psychic voyage. As a kind of focus, so to speak."

Mr. Axelrod grew pensive for a moment. "There would seem to be more to it than that. Surely you've thought about it?"

I hesitated. "It's a bit difficult to articulate." Axelrod brightened up. "Try me."

"Well, I have to introduce the possibility of.... Well, we are educated to believe that thought takes place only inside of one's head, in the brain—that the mind is inside each person's head.

"But this runs counter to the fact that some things can be directly shared at a group level—maybe not thought itself, but certainly emotions and sentiments, for example."

"For example?" Axelrod asked.

"Well, during the 1930s, a lot of work was done on what was called 'mob consciousness', where anger or hysteria seems to get communicated by means other than reason or logic. This was suggestive of a group-mind kind of thing—somewhat linked together by a kind of communal telepathy.

"In the middle ages there were lots of communal phenomena, or hysterias like this..."

With this, I thought I noticed some kind of change in Axelrod—a slight pink color in his face. One can tell if someone is accepting or resisting. It's a sort of well-known magnetic thing.

I went on. "If there is a group mind, there possibly could be a species-wide one—having some kind of memory...which individuals could link into..."

Axelrod interrupted. "Are you speaking about some kind of Akashic record or something like that?" He now DID seem nervous.

"No, not exactly. Some kind *of* species memory storage—maybe at the DNA molecular level. I know this idea makes scientists throw up, but so does any aspect of Psi."

I paused to see if this passed inspection by Mr. Axelrod. He was very quiet and so I couldn't tell. But finally, he said, "continue."

"There has been a lot of interest as to why the coordinate thing should work. I've discussed it with Dr. Jacques Vallee, the famous Ufologist, along the lines of information theory.

"Certain theories regarding information hold that it exists everywhere as sort of a cosmic thing. And if one had an 'address' for it, one could link into it, like a computer that can find information if there is a correct address for it."

"Are you suggesting," Axelrod asked, "that the mind is a computer that can link into..."

"Well, something like that, but not at the intellectual level. There actually must be lots of mind layers that function differently."

"But why should coordinates function...?" Axelrod mused, almost to himself.

"Well, in a cosmic sense, if one has a round ball or a planet, and if one wants to divide it up, one will assign what amount to longitudes and latitudes. These will divide the ball into segments. If Intelligence exists as a universal, then this would be the best universal way to divide up a planet so as to know where one is on it."

"This is a matter of triangulation. Is this not how illegal radios are found—by sending out two or three cars with antennae that can get a triangulated fix. I saw this in a World War II movie."

"Universal?" Axelrod asked. Now a "hot" magnetic thing seemed to come out of him. "Why did you use that word?"

Well, why NOT I thought. "Well, the best evidence we have for telepathy, for example, is that it seems to be universal to our species. People experience it regardless of their different cultures, their different backgrounds. If we assume that Intelligence can be universal, we also have to assume that Intelligence also has to have sensing factors that also are universal."

Having said my piece, I awaited Axelrod's comment. He just sat looking at me in an odd kind of way. Suddenly I got the idea: Aha, he has some coordinates in the Soviet Union he wants me to look at. After all, everyone else had them.

However, Axel now resumed his smile. "But you went to the planet Jupiter. Did you use coordinates to do so?"

"Well, yes and no. The Jupiter thing came about as a sort of a lark. Again, as at the American Society for Psychical Research in New York, at SRI I got bored with the hundreds and hundreds of experiments.

"NASA was sending the Pioneer fly-by past Jupiter, and I thought it would break the monotony of our SRI work by trying to get to Jupiter ahead of the fly-by. It was a good experimental idea, for we could register my impressions of the planet, circulate them to interested people, and do it in advance of the data the fly-by would send back.

"This data would act as feedback to see if we nailed down any unsuspected facts about Jupiter. It was just a further test of remote-viewing capabilities.

"As kind of coordinates, we found out where Jupiter was in relation to which part of the Zodiac, where Earth was in respect to the Sun. These three factors—the placements of Earth, Sun and Jupiter, acted as a kind of triangulation as to where Jupiter was."

"Yes, I see," Axel grinned. "You did pretty good."

I decided to take an initiative. "Axel, I don't like to do tasks unless there is a good chance of obtaining feedback, and you represent one of those times I have been dragged into a situation where obviously I am not going to get any...am I?"

"Well, that poses a bit of a problem considering our situation here. But surprisingly some feedback will become available in other ways. I'll send it to you, in an unmarked envelope."

"Well, what then is your task?" I asked.

After a long moment, Axel asked: "Ingo, what do you know about the Moon?"

The MOON! He wants me to go to the Moon. "Well, I know it is there, that it's a dead satellite, it has craters and mountains, if that is what you mean."

"Have you studied the Moon, or gone psychically to it?"

"No. We never tried the Moon, because too much is known about it. It would not constitute a good experiment because of that. People would think I had learned about the Moon or looked at it through telescopes or something."

"What about the reverse side of the Moon. That side is always turned away from Earth. No one could accuse you of being able visually to see that."

"But, still, the NASA's Moon missions have encircled it, and there are lots of photos and stuff."

Axel laughed. "Well, we want you to go to the Moon for us and describe what you see. I have some Moon coordinates prepared, about ten altogether. Is that too many?"

"Well, no, depending on stress factors. But I don't like to do too many at once, because I fear I will begin to superimpose my impressions."

"Well, we may not have to do all of them," Axel said cryptically. "Do you know who George Leonard is, or ever heard of him?"

"No."

"You're quite sure?"

"Well, I've met hundreds of people by now, but I don't recall any George Leonard. There's a Leonard at SRI, but I can't remember names very well, anyway. Faces I remember better."

Axel immediately fumbled in a folder, pulling out five photographs. "Are any of these familiar?"

"Well, one is Dr. Karlis Osis, and this other one works out at SRI, but I don't know his name. I've never seen the other three, one of whom I suppose is your Mr. Leonard."

"Well, good, then. We seem to be in good shape. Now, how about a work-out in the gym, and then I'll join you for dinner. We can start early tomorrow morning."

So, the initial interview was over. I am not an exercise buff, but I wanted to go to the gym hoping to see more people and more of this astonishing underground facility.

I was to be disappointed.

The twins accompanied me along empty corridors to the locker room, and themselves got geared for exercise. Built like brick shit-houses they were, and took turns doing hundreds of rapid push-ups, making me feel undeveloped, as I was, when it comes to physical strength and stamina.

But they SPOKE every now and then—"Mr. Swann, that weight might be too heavy for you."

NOW I could distinguish a DIFFERENCE between them. One had a Southern drawl, while the other had what I thought was Australian one, of all things.

This set me wondering.

Why, for example, if all this was ultra-secretive, had I been met by two men who were so obviously twins, and extraordinarily handsome. Surely this would have attracted attention in the museum's rotunda. But then I remembered it had not. And then I remembered, too, that most people notice very little to begin with.

Gradually, I began to realize that the two actually did not look alike; they only SEEMED to in some inexplicable way.

Suddenly I could see great differences between them. Their physiques were almost identical, but I eventually noticed that one was slightly smaller.

One, the Australian, was older. Their square jaws and green eyes were alike, but the noses were different, and one had narrower lips. Their hair was identically close-shaven in the well-known military style, but now I felt they were different.

So, they were not twins, after all. Nor even brothers, doubtlessly. But what was it about them made them so alike, as if to be mistaken for twins?

Their energies! Something about their energies.

One of the basic characteristics of good psychics is their fascination of observing everything they can, and in detail. Powers of observing seem to act as a launch pad for higher forms of perception. I'd had this fascination from childhood.

As I watched them more carefully, I slowly became conscious that they MOVED almost as if in unison. If one lifted a hand, so did the other.

They moved almost as if of one mind, so to speak. Yes, that was it. They were enough alike so as to be mistaken for mirror images of each other—until they spoke, that is.

The word "entrainment" came to mind, a word used to describe people who have been subjected to some mind-managing so that they begin to think, act, and even, I guess, look alike.

I began to get the fanciful idea the twins were cyborgs or androids of some kind, but then decided my imagination had taken over.

Needless to say, I never found out what the twins were or why they were so un-twin-like, yet so alike.

The twins and I swam a few laps in the small pool in the underground installation, they swimming most of the time underwater.

Coming back from the pool, I found the jolly Mr. Axelrod standing near a small table loaded with food. We ate a great steak dinner with all the trimmings, except that I couldn't drink the obviously good wine because I was going to "work" in the morning.

We had a somewhat cheery conversation while eating.

Among other topics brought up by Axelrod, he wanted to know more about what I knew about telepathy. We chatted this up. I thought this was merely innocent conversation.

PSYCHIC TOUCHDOWN ON THE MOON

I had spent a nervous night. First of all, the bed was kind of hard, and there was no sound in the room. So I listened to my heart pumping in the dark silence. I felt a little claustrophobic which reminded me of how I had responded inside the Great Pyramid in Egypt when I had visited it in 1973.

I sifted through possibilities—wondering if all of this could possibly be a deeply covert KGB thing. Axelrod LOOKED and ACTED American enough. But the twins?

Openly confessed, my attention was on the $1,000-day thing. By 1975, I had been in Psi research for about five years.

When it was arranged for me to take part in those earlier experiments, the first order of business in the minds of the researchers was to figure out how to pay me the least possible, and preferably to pay me nothing.

The $1,000-day was a real, and much needed windfall for me. So I worried about the many ways it could get messed up.

Failing to provide good Psi data was one way. But, as I had found, if one talks about things people don't understand, then they lose interest. Another way was NOT to provide what the client wanted.

I had no idea what Axelrod wanted. Maybe they, whomever THEY were, were looking for good places to build Moonbases. Maybe THEY had lost a secret spacecraft or something along those lines.

But there I was, nervous or not, deep underground somewhere, twisting on a hard bed. Well, I'd remote view the Moon and get it over with. I didn't expect to see much on the Moondead satellite, airless, dust, craters, etc.

Anyhow, the mystery WAS a bit much if one took time to think about it. Being dragged around in hoods! Really! I determined never to get caught up in such an affair again.

We started our work early the next morning—which I immediately dubbed "Moon Probe."

As we had done in the Jupiter project, I asked Axel to find out where the Moon was in its monthly cycle—which is to say, its present relationship to the Earth and the Sun.

"The Moon is full," he began, "opposite the Sun, and the Moon is just setting in the West. Will that do?"

"I hope so," I replied. "Earth is between the Sun and Moon, then, and what I have to try to do is head directly away from the Sun, hoping for a psychic touchdown (I smiled in saying this) on the Moon surface."

"Ok, do your thing, then," Axel smiled. He pushed the "record" button on his tape recorder.

Earlier that morning, we had discussed the experiment's protocol, the way the session was to be conducted.

Except for voicing the lunar coordinates when I asked for them, Axel was to make no verbal inputs.

I talk out loud when "doing my thing," asking MYSELF a series of questions. But these are questions to aid my intellect in trying to understand what I am experiencing. These are not questions others need to respond to while I am "at work." I DO NOT like to close my eyes when I am doing my "thing."

I settled back and tried to get a sense of Earth being between the Sun and Moon—and slowly began to have images of rising upwards from Earth until I could see its curvature.

As I had already learned from our efforts to get psychically to Jupiter, the Sun looks much smaller to my psychic senses than it does when we eye-ball it on Earth. Seen psychically, and if seeming smaller, at least three "envelopes" of some kind are clearly visible around the solar star.

In any event, I tried psychically to head away from the Sun, toward the Moon. This now looked LARGER than it does when eyeballing it.

I had no problem getting there. Slowly at first, it grew larger and larger and then swiftly filled my psychic vision completely—a whitish thing, with grays, darks and, surprisingly, a lot of yellows in it. Suddenly I was kind of sucked into-toward it faster, as if in a gravity free-fall. Next, I had the sensation of "being" next to some pumice-like rocks.

"OK," I whispered to Axel, "I can see these rocks, and some dust, so I guess I must be here. Give me your first Moon coordinate preceding it with the word Moon."

I wrote down "Moon" and the coordinate-ordinate and nothing happened. I was still where I had touched down.

"Give it again, more slowly," I asked. He did so, and I experienced a blurred kind of vision, a sense of zooming across a plain, some mountains—and then into darkness which surprised me.

"It's dark here," I said. "Why is that? A rhetorical question, Axel. Please do not speak an answer." Darkness! Then, slowly, as if adjusting to a kind of night vision, I could begin to perceive formations. And I realized what had happened.

"This coordinate," I asked, "is it on the dark side of the Moon? Yes, it must be."

I began trying to make sense of the impressions I was acquiring. "Well, I seem to be near a cliff of some kind. It goes upward quite high, made of some kind of dark rock. There is whitish sand, a fluffy kind of sand. Away from the cliff formation there is a broad expanse of some kind. There are some patterns in the sand, or whatever it is—not quite like sand."

"What do the patterns look like?" Axel interjected. He was not supposed to intrude with questions. But he had, so I went with it.

"Well (I now closed my eyes), sort of like little tufts or dunes, as if the wind had made a kind of pattern."

After a moment of considering these little dunes: "But there is not supposed to be any wind on the Moon, is there? No atmosphere? Yet, I can sense something like atmosphere. I'm getting a little confused. Let's take a break."

Was I mistaken? Axelrod seemed to be looking at me in a rather strange way—as if swallowing a desire to speak.

"Well," I went on, "what they actually look like are like rows of largish tractor tread marks. But I don't understand how this could be, so they must be something I don't understand. They are just marks of some kind. Strange, though."

I was silent for a moment. "Axel, do you want—well, am I supposed to try to see metals or something here, or what? I'm just near this cliff here—it has a kind of shiny quality to it, something like obsidian..."

Axel answered: "No, we can go on to the next coordinate-ordinate now."

"Give me a moment," I asked, "then on my signal lay it on me."

I wrote down the next coordinate-ordinate. The cliff vision faded, and in a few moments, I was clearly at another place, which I could hardly believe was on the Moon. "I'm sorry, Axel, I seem to have gotten back to Earth here..."

"Why do you think that?" he asked.

"Well, there are...some..." I stopped. I looked at Axel. "Maybe we better take a break, a little coffee, and then we can try again."

"OK, but what did you see?"

"I have no idea. But whatever it was it couldn't be on the Moon." (I had visions of $1,000-days coming to an abrupt end.)

So we had coffee and chatted up this and that. Axel, for the first time, seemed somewhat nervous.

In about fifteen minutes we got back to it. I went through the same process of going away from the Sun until I was at the Moon. "OK, give me that coordinate-ordinate again."

He did. I wrote it slowly down, making sure I made no error in doing so. I became aware of a greenish haze, which is what I had seen before. This time I decided to go for it, for better or worse.

"Well, I am in a place which is sort of down, like in a crater I suppose. There is this strange green haze, like a light of some kind. Beyond that, all around is dark though. I am wondering where the light is coming from..." I jolted to a stop again.

After a moment, Axel prodded. "Yes, what else?"

"Well, you won't like this, I guess. I see, or at least think I see, well...some actual lights. They are giving off a green light...I see two rows of them, yes, sort of like lights at football arenas, high up, banks of them.

"Up on towers of some kind..." I gave up here. "Well, Axel, I can't be on the Moon. I guess I have to apologize, I seem to be getting somewhere here on Earth."

Axel stared at me for a moment. He was NOT smiling or looking sympathetic or tolerant. I thought it was all over with. "You're sure you see lights, actual lights?" he finally asked.

"Well, I see lights! But how can they be on the Moon?"

Axel had a pencil in his hands, which he was twirling around and around. His not-smile developed into a frown. "Shit," he finally uttered, and broke the pencil in half. I was quite surprised and fully expected him to stand up and leave the room in dismay at my remote-viewing flub. But he did not.

"Lights, huh? You are sure you saw lights?"

"Well, yes. But not on the Moon, surely. How could they be on the Moon?"

Axel stared at me, saying nothing.

I can be quite dense, I suppose, but something started twanging around in the denseness.

I blinked at Axel. "You mean..." I began, somewhat uncertain as to what to say. I realized I had to select my words carefully. "Am I to think these lights are actually on the Moon?"

There was no answer forthcoming from Axel. I pressed onward. "Have the Russians built a Moonbase or something? Is that what I am supposed to be remote-viewing?" Again, no answer.

We sat and stared at each other for a longish period, he not willing to commit. After a moment of this confrontation, I decided to reassume the initiative. "Maybe you should give me that coordinate-ordinate again."

Once back in the glow of the greenish lights, I now seemed to have the courage to begin really looking. "Well, the light seems diffused somehow, as if there is a lot of fog—no, it's dust—dust! Floating in the air."

I paused, then continued: "Yet there is no air on the Moon, is there? There is noise of some kind, like a thumping. I can see one of the light towers better now. Hey, it seems built of some very narrow struts of some kind, thin like pencils. Like some sort of pre-fab stuff right out of Buckminster Fuller's stuff."

"How high are the light towers?" Axel interrupted.

"Well, high enough. I have to find something against which to compare them. Let's see...hey, there are some of those tractor-tread marks everywhere. If I guess these are about a foot wide, well, then, let's see, if I compute as correctly as I can, well..." I paused, looking at Axel.

He was not smiling. "Yes?" he arched his eyebrows.

"Well, tall...about or let's say over a hundred feet. But?"

"But what," Axel asked, leaning forward.

I swallowed hard, and almost chickened out at this point. "Well, I think I got a glimpse of the crater's edge. On it I think I saw a very large tower, very high that is."

"Yes?"

"Yes! *Big,* really big."

"How big?"

I swallowed again. "Well, if I compare it to something I am familiar with in New York, about as high as the Secretariat building at the United Nations—which has thirty-nine floors in it."

Axel narrowed his lips. "You can see that, then?"

But this, as I took it, was a question Axel was asking himself more than me.

Again the silence. I decided to again assume whatever initiative I could.

"Am I, then, to assume this stuff really IS on the Moon? If so, this is more than a Moon base, isn't it, Axel?"

Again no answer. So I continued: "But this stuff is big. Does NASA or the Soviet space program have the capabilities of getting such large stuff onto the Moon? I thought everyone was having trouble just getting a couple of guys and a dog into orbit. I thought the only thing we got on the Moon was a flag planted in some crater somewhere."

As I talked myself through all this a certain glimmer began to dawn in the recesses of my mental darkness. I suddenly stopped speaking.

I stared incredulously at Axel. "You mean—am I to assume this stuff is—not OURS! Not made on Earth?"

Axel raised his eyebrows, trying to grin. "Quite a surprise, isn't it," he said. I had a sense he was trying NOT to be emotional.

Surprise? To say the least! I was completely dumbfounded to the degree that I had begun taking very short breaths, getting dizzy thereby. "I take it you would like a break before we continue," Axel ventured.

What I really needed was a recovery couch. In fact, I still get breathless even as I write these very words now.

It's one thing to read about UFOs and stuff in the papers or in books. It is another to hear rumors about the military or government having an interest in such matters, rumors which say they have captured aliens and downed alien space craft.

But it's quite another matter to find oneself in a situation which obviously confirms EVERYTHING. Not principally because I suddenly knew the rumors were true. But because I found myself in a situation in which I, in my psychic processes, had seen the evidence for myself.

"Good heavens!" I breathed.

My brains began racing, putting things together. Axel, though, and the twins, and the elaborate secrecy of this whole "mission," was the best evidence.

I was now completely certain that I was physically present in some kind of ultra-ultra-ultra-secret place, and that the mission of this place was to sort out extraterrestrial matters.

I knew that NASA must have photographic evidence of activities on the Moon which already confirmed the presence of extraterrestrial activity there.

What I did not understand though, and I realized this only with my third cup of coffee and my tenth cigar, was why this ultra-secret project needed my services.

So, I looked at Axel, and this time I was not smiling either. "Why the hell have you dragged me into this, Axel? If you possess enough to cause what I take is your mission to come into existence, surely you don't need my inputs here?"

"Well, Ingo, no—and yes."

"I'm confused," I said, rather sternly. "Please explain."

"I can't. Well, at least I cannot give you information. It was felt that doing so would jeopardize not only us, but our mission. You seem bright enough, though."

"Thanks heaps, Axel. Well, this bit of security almost blew it for you. Had I not learned some time ago to accept and describe what

I was seeing psychically, accept it BEFORE prejudging it, I would have not dared to say I saw lights on the Moon. I would have edited that out, fearing others might see me as loony. God damn it! ETs on the Moon, no less!"

"Well," Axelrod began, "if I had told you in advance, would you have thought I was loony?" Axel asked.

He had a point.

AND ALL OF THIS WAS REAL! I closed my eyes as waves of goosebumps cascaded through my body. I couldn't control them, so I broke into tears.

"Shall I leave you to recover?" Axel asked.

"If I can be by myself yes, but if one of those god-damned muscle men twins has to stand and watch me blubber, don't you dare. No way do I want THEM to see me in this condition."

"They would understand perfectly. We all have experienced a considerable amount of emotional surprise."

"I can't believe that either of those twins would ever think about crying...". But suddenly, through my emotions I started laughing, almost uncontrollably. "This is SERIOUS, isn't it?" I finally managed to blurt out. But my thoughts were going a mile a minute.

The bottom line: We are not alone—and some ultra-secret, presumably a governmental agency, goddamned well knew it! My glee changed swiftly into anger. Shit! Shit-shit-shit!

"Well," I snarled, "whoever is in charge of these matters hasn't managed them very well as far as us ordinary public types are concerned."

"I'll concede that, Ingo," Axel said. "Frankly, no one has known what to do, and many mistakes have been made."

"Yes, and all in the name of what—privileged information in favor of the few, of the military, of scientists, or what?"

"Sometimes. But the problems are more than you can imagine."

"Don't give me that, Axel. Here you drag me into a very scary situation, ask me to utilize my thing in a very strange way, and ask me to see FOR YOU something I cannot imagine?

"Get outahere? I don't buy it. I don't like this; I don't like it at all."

Axel and I sat staring at each other. Neither of us was smiling. "Do you want to leave, then?" he finally asked. "We will do whatever you want."

Of course I did NOT want to LEAVE! I wanted to understand. "Why do you need my services, Axel? Just answer that one question.

"If that stuff is on the Moon, why don't you just send along another Moon mission to have a good look-see...?"

But the awful truth dawned in a burst of light.

I looked at him. "Unless they...I can't believe this, unless they somehow have told you to stay away, and somehow shown you they mean it!"

This time Axel was neither smiling nor not smiling. I got out of my chair and started pacing the length of the table.

I started laughing. "Goddamn it! They've somehow got you by the balls, haven't they. That's why you are resorting to psychic perceptions! Ja-eesus Kahariiiist!? They are NOT friendly, are they? ARE THEY, Axel!"

Axel kept his cool. "There are two major reasons why I've asked you to help. You are approximately correct about the first one, but not completely so. The second reason is more simple.

"Your information might provide a kind of check point in what you surely now realize must be a mess of interpretations of the photographic and other evidence.

"It was my idea to find a psychic who did not know anything about the Moon and see what might be seen there. Sort of an independent source *of* information, which would lean our interpretations one way or another."

"Have you used other psychics, then?" I demanded, very intent on getting an answer.

"Please don't require me to say yes or no."

I felt my patience thinning out. "Why not?"

"There are several reasons, but mainly there is a confidence factor involved."

"Confidence about what, about the abilities of other psychics?"

"Yes, that is one of them."

I sat back in my chair and trying to keep my hands from shaking too obviously, lit another cigar. My brain was whirling.

"So," I began, "the only major way now to spy on those guys is to resort to psychic abilities, which the mainstream of our great nation makes a special effort to discredit. What a gas! What a complete gas!"

I started giggling. "Well, speaking of who's got who by the balls, I've suddenly got you by yours haven't I?"

Axel sighed. "Well, they said you were quick on the uptake, and stubborn, and could throw tantrums. I see they were right."

"'They', who is 'they'?" I asked, but I couldn't stop giggling.

Before he could answer, which he obviously didn't want to anyway, I had another brainstorm coming on.

"I suppose, then, the Soviets are having the same problem. Don't tell me THE SOVIETS have resorted to THEIR psychics!!!?"

Axel had resumed his stoic smiling-not-smiling face. I jumped out of my chair again. "Got you again, haven't I?" I almost shouted. "You KNOW the Russians are using psychics, and you are afraid they will get psychic Moon-information before we do! I'll be fucked!"

At this point, and since I now felt I had a grip on everything, I suddenly felt energetically depleted. "I want to take a twenty-minute nap," I stated, and headed for the bedroom. "After that and some food, we can get back to work."

I don't remember even getting onto the bed, but I suppose I did, and afterward learned I had slept six hours.

HUMANOIDS ON THE MOON

B ack at work, Axel gave me Moon coordinates, each set representing specific locations on the Moon's surface. At some of the locations there seemed to be nothing to see except Moonscapes.

But at other locations?—well, there were confusions, and I perceived a lot that I could not understand at all. I made a lot of sketches, identifying them as this or that, or looking like something else. Without comments, Axelrod quickly took possession of each sketch, and I was never to see them again.

I found towers, machinery, lights of different colors, strange-looking "buildings."

I found bridges whose function I couldn't figure out. One of them just arched out—and never landed anywhere. There were a lot of domes of various sizes, round things, things like small saucers with windows. These were stored next to crater sides, sometimes in caves, sometimes in what looked like airfield hangars.

I had problems estimating sizes. But some of the "things" were very large.

I found long tube-like things, machinery-tractor-like things going up and down hills, straight roads extending some miles, obelisks which had no apparent function.

There were large platforms on domes, large cross-like structures.

Holes being dug into crater walls and floors obviously having to do with some kind of mining or earth-moving operations.

There were "nets" over craters, "houses" in which someone obviously lived, except that I couldn't see who—save in one case.

In THAT case, I saw some kind of people busy at work on something I could not figure out. The place was dark. The "air" was filled with a fine dust, and there was some kind of illumination—like a dark lime-green fog or mist.

The thing about them was that they either were human or looked exactly like us—but they were all males, as I could well see since they were all butt-ass naked. I had absolutely no idea why. They seemed to be digging into a hillside or a cliff.

As I described, "They must have some way of creating a good environment, warm and with air in it. But why would they be going around naked?" No answer was forthcoming to this self-question.

But being there in my psychic state, as I felt I was, some of those guys started talking excitedly and gesticulating. Two of them pointed in my "direction."

Immediately I felt like "running away" and hiding, which I guess I, psychically, did, since I "lost" sight of this particular imaging.

"I think they have spotted me, Axel. They were pointing at me I think. How could they do that...unless...they have some kind of high psychic perceptions, too?"

Axel said, in a calm, low voice, so low I hardly heard it at first. "Please quickly come away from that place."

My eyes were wide as understanding drained in. "You already know they are psychic, don't you?" Axel raised his eyebrows and gave a deep sigh.

And, at that point, he abruptly closed his folders. "I think we had better end our work here."

I was quite surprised. But I had not fallen off the psychic truck just yesterday. "You think, you already KNOW, that they have some kind of, uh, telepathy-that they can trace where this psychic probe is coming from? Is that it?"

Axel had started smiling again, but obviously was not going to respond. "Come on, Axel, loosen up a little."

But I was not to be deterred. "Would they kill an Earth-psychic if they felt he or she was good enough to spy on them?"

"There is no conclusive evidence to suggest that," Axel responded.

I gritted my teeth. "No 'conclusive' evidence! What the hell does THAT mean?" My voice had climbed several octaves.

"It's very difficult for us to assess any of this," Axelrod began. "We don't know, but that they do have things and capabilities we here are trying to understand is very apparent. Whether they spotted you or not will be unclear, but we have to put no prejudgments on what guides our mission.

"At any rate, we don't want to put you to any more risk. Let's eat some dinner, and then get you back to New York.

"I'm afraid we have to repeat the process used to get you out of here. I hope you don't mind. We are very grateful."

"RISK!!! What do you mean by RISK?"

I could see that Axelrod was prepared to be noncommittal. So I took the initiative. "If it's telepathy, then it's a different kind, at least from how it is understood here on Earth. It's NOT just telepathy, either."

THIS got his attention. He looked at me in surprise. "What do you mean?"

At this, I FINALLY comprehended that his earlier interest in telepathy had not been just innocent chit-chat.

"Well, I don't know exactly. It's more than just mind-to-mind. It's like, well...". I was grasping for words. "Well, when they 'saw' me, they couldn't really see me, could they? What, then were they seeing? I'm asking myself this, Axel?"

"Yes, go on," he said.

"Well, it's more like they were...FEELING rather than seeing or picking up on mind vibes. It's more like it was, yes, sort of a dimensional thing—rather, sort of like a ripple in some kind of cross-dimensionality. Yes! That's it! They FELT something. Not particularly ME. But SOMETHING."

I paused: "And! THEY knew what the ripple meant. Like a sort of penetration of where they were." I paused, then said in a self-introspecting way: "WOW!"

Axelrod sat quietly, as was his way, looking at me. Then: "Why did you say WOW?"

"Ah! Well, if I can articulate it...it was like there is a sort of... cross-dimensional....Well, if you can imagine that you feel a presence but can't see it, it was something like that.

"Only those guys...they were going to hone in on it, at least that's the best way to describe it."

Axelrod was silent for a moment. "So, you are referring to telepathy plus something else?"

"No, not exactly. Perhaps SOMETHING ELSE plus telepathy. It's the other way around. After all, the basis for telepathy has to come from something—rather, because of something."

"What do you mean?"

"Well, nothing happens all by itself. There are always processes involved. This is to say, things happen by way of something. Nothing comes out of thin air. It's hard to articulate in simple three-dimensional terms. Mind-to-mind is a three-dimensional construct. But what if..."

Axelrod interrupted. "Why would mind-to-mind be a three dimensional...?"

"Well, one mind existing as a three-dimensional thing communicating with another which is also a three-dimensional thing, and the communicating across the distance...is not the distance involved conceived of in three-dimensional terms?

"The PHYSICAL universe is three-dimensional, not the mental universe. Here is one of the big flaws in all theories about Psi. Everyone thinks of Psi ONLY in three-dimensional terms." I ran out of words at this point.

Axelrod was looking at me with his calm, unblinking eyes. But he was lightly drumming his fingers on the table. I knew I had hit something of concern. And his next comment proved it.

"Could you write down your ideas along these lines?" I could. I did. I remember producing fifteen hand-written pages.

After this somewhat inscrutable advisory, there came a handshake, the hood, a chopper ride, and by the twins and the same car I was delivered back to the center of Washington and let off at the train station at my request. The twins said no more than they had to. I found myself wondering if they actually came from the Moon.

I spent the next few months wondering if the ETs were going to find me and zap my brains out of existence.

When I left Mr. Axelrod's carefully hidden establishment, he reminded me of my pledged ten years of confidentiality.

"Not to worry, Axel," I replied. "I have no intention of demolishing my official research work by introducing something so far out as claiming I have seen extraterrestrials working away on the

Moon. No one would believe me anyway."

I have abided by that promise, well past the ten-year mark. The reasons I have now decided to write about all this will become clear in later chapters.

As I departed, Mr. Axelrod asked that if he again had need of me, would I be interested. "Probably," I responded, for how could I not be—Jesus Christ, ETs on the Moon and some official investigative agency? Who could resist!

"Good," he replied. "But my name Axelrod is now retired when you leave here and will not be used again. We will be in touch with you in some other fashion, which I will make sure you recognize.

"If anyone ever asks you about 'Mr. Axelrod' or about this place, or asks if you know anything about it, such inquiries will not be coming from us. Please act accordingly, for our sakes and your own."

God! Scary, huh? What had I gotten into? But his advice came in handy when, about three years later, my telephone rang.

It was a Mr. Dillins or Dallons (I didn't quite get which) who said he was an investigative reporter digging into government cover-ups of the UFO situation.

I said I didn't know anything about that—other than what I read in various books and articles. He brushed aside my evasion and asked if I knew Mr. Axelrod.

"Who?" I asked in return.

"You know," the investigative reporter said, "Mr. Axelrod."

"Never heard of him," I replied.

There was a silence at the other end of the telephone, and then the caller clicked off without so much as a thank you or good-bye—leaving me with shaking hands and much in memory of Axel's forewarning. You want a basis for paranoiac tensions?

After leaving Mr. Axelrod, and back in New York, I decided I was pretty much of a wreck. I slept for about two days, dumbly watching the boob-tube between naps and such. I ate a lot.

Then, when I began to get IT and all ITS implications back together, I decided to make some sketches of what I remember drawing for Axel.

I couldn't remember any of the coordinates-ordinate numbers, and the names of the plains and craters on the Moon were never used when we were doing the remote viewings.

So I don't know where these intelligence-made structures and such were actually located on the Moon. But I could hardly forget what I had seen.

I made several larger drawings, and then decided to fit them onto two pages—which I then placed in my bank's safe box, since I had visions of my home and studio covertly being gone through without my knowledge.

Paranoia rode high with me for quite some time. But I suppose my safe box was just as accessible to the covert powers that be. There were two pages of the sketches, and which will be presented ahead.

FEEDBACK
[OF SORTS]

Considering the rather dramatic aspects of the Axelrod affair, it might at first seem unlikely that I could forget about it. But except for meeting the contacts in front of a stuffed elephant and having a hood placed over my head, the Axelrod affair wasn't all that different from many other official and unofficial experiments I got involved with.

Many of those experiments had equally dramatic elements. Most were done in careful secrecy, and my weekly schedule along these lines was jammed with this kind of activity.

One might wonder how it is possible to forget about humanoids and structures on the Moon.

Well, for one thing, there was a 50-50 chance they were there or not there. Further, as with all Psi experiments, there was a 100 per cent chance I had been viewing my imagination and fantasies.

As I took it, there was to be no feedback to help resolve the imagination issue one way or another. Since people tend to operate based on feedback, they tend to forget about stuff that never achieves the feedback.

Then there are two additional, but extremely subtle phenomena that appear to be involved. They can begin to take on meaning only if they are identified and opened up for inspection.

The first phenomenon has to do with the fact that most people forget (and avoid) whatever does not fit within consensus realities. The second phenomenon has to do with the fact that most humans forget about the Moon altogether. It is THERE, of course. But beyond that, interest in it is exceedingly minimal.

It is somewhat difficult to articulate this. One way of beginning to do so is to point up that people are very much interested in Mars, for example, or in the possible existence of Intelligent Beings somewhere in the VERY far distant reaches of space.

But with the Moon, it's almost as if Earthside human

consciousness of the satellite is somehow rigged so as (1) to avoid thinking too much about it; and (2) to disregard any unusual lunar phenomena. I'll try better to articulate these two factors ahead.

As it was back in 1975 and 1976, if not actually forgotten, my memory of the Axelrod affair had receded into some deeper subliminal areas of memory storage. And if I thought about the affair at all, it was only to note that it had happened, that it was over and done with, and that I didn't dare to talk about it for reasons so numerous that it was the better part of valor to forget them, too.

However, whoever or whatever is behind the scenes dealing out the cards of circumstances was not finished with the Axelrod affair.

For at some point in 1976 (during the summer, I think), what might be called the second chapter of the Axelrod affair opened up.

For I received in the mail, a plain envelope which did not bear a return address or even a postmark, although it did have stamps.

The envelope contained a book and nothing more.

It was entitled *Somebody Else Is on the Moon*. The author's name was George Leonard. I spent the next few hours reading it, and then re-read it two more times.

Apparently, at the time of my ultra-secret visit, Mr. Axelrod had already known that this book was coming out, and of course he had been interested in whether I knew the author or not.

Leonard apparently had obtained NASA photographs *of* the Moon, which after all are in the public domain because most of NASA's work is funded by our tax-paying money.

"What NASA knows," the frontispiece of the book began, "but won't divulge! With careful logic and reason, George Leonard has studied all the data (including official NASA photographs and the astronaut's Apollo tapes) to prove his theory of a highly advanced underground civilization that is working the surface of the Moon— mining, manufacturing, communicating, and building!"

Leonard's book was filled with verifiable data, official photos, and sketches of structures etc., he created from the photos.

Well, I can tell you I ran to my own drawings and spent a week comparing and re-comparing them to the sketches and photos George Leonard had provided in his own book. Many of Leonard's sketches resembled some of mine.

Yes indeed, the mysterious Mr. Axelrod had provided me feedback as promised, for there could be no doubt that it was his jolly self that sent Leonard's book to me.

But could Leonard's book be considered adequate feedback?

Well, if not completely, at least somewhat.

For example, ARE there structures on the Moon?

As Leonard pointed up, one of the most remarkable photos, taken by the astronauts of Apollo 12 on their flights around the Moon, portrays what is referred to as Super Rig 1971 (NASA photo 71-H-781) and which is very similar to another photo of a similar Super Rig (NASA photo 66-H-1293) taken five years earlier.

The astronauts of Apollo 14 (1971) obviously were EXPECTING to see this Rig or one like it. When it apparently came into view, they referred to it as "Annabel"—which was "just like the one we saw yesterday. She's sitting right on the ledge and must be over a mile high. Did you see THAT! The light flares coming from the dark part of the crater, just below Annabel. Oh, cameras, don't fail us now!" (This conversation is paraphrased from that given on page 54 of Leonard's book.)

Indeed, there seem to be a number of "towers" on the Moon, and an equal amount of confusion about them. As I discovered (when I later began my Moon research in earnest), during the early 1960s NASA sent Orbiters to the Moon in preparation for the manned Apollo missions.

A released NASA photo numbered Lunar Orbiter III-84M quite clearly shows two structures rising up in the Sinus Medii region.

The first of these became referred to as "The Shard." This structure towers up from the Moon's surface for about a mile and a half. Near "The Shard" is another structure referred to as "The Tower." This has been photographed four times from two different altitudes. It rises up about five miles and is capped with what appears to be cubes joined together to form a very large mushroom-like crown having an estimated width of over a mile.

Several independent geologists who have examined the photos indicate that no known natural process can explain the two structures, which is something of an understatement.

Please note that copies of the photos referred to above can still be obtained from NASA supplier. But I've been told that evidence of the towers has by now been airbrushed out.

After studying Leonard's book, during the next two weeks I wasn't certain whether to sleep or stay awake, and all my bio and mental cycles found themselves quite interrupted.

I fully expected that Leonard's book would shake the nerves of all us Earthlings.

But, most people I ranted to just smiled and said that Leonard's opinions must be just that, and that "there must be some other and more logical explanation."

Even some UFO people I knew were hardly interested, a factor which I found (then and now) quite confusing and mysterious.

As it turned out, it seemed that most people simply couldn't handle the implications of Leonard's book. Today when I mention it to people, well, they blink, have never heard of it and a sort of film appears on their eyeballs. That they don't WANT to hear of it is more likely the case.

Well, I was interested by the implications. For if there are extraterrestrials on the Moon, surely getting themselves to Earth should not be much of a problem for them.

I thus decided, in my small mental recesses to be sure, that we on Earth may have neighbors who are not from Earth!

And/or that some of our systems and organizations may indeed "contain" extraterrestrial "influences" in them.

I mean, how much does it take to put two and two together?

I began to see why Mr. Axelrod's group, if they themselves were not extraterrestrials (as I wondered at times) was ultra-secret and came and went under changing subterfuges of one kind or another.

Soon, however, I was sucked back into the hectic pace of my life and research work. Soon, I'd "forgotten" all of this. If I thought about it at all, it was only to have the fleck of an idea that the Leonard book constituted the feedback, and so that was that.

Ingo Swann Moon Sketches
March 1975

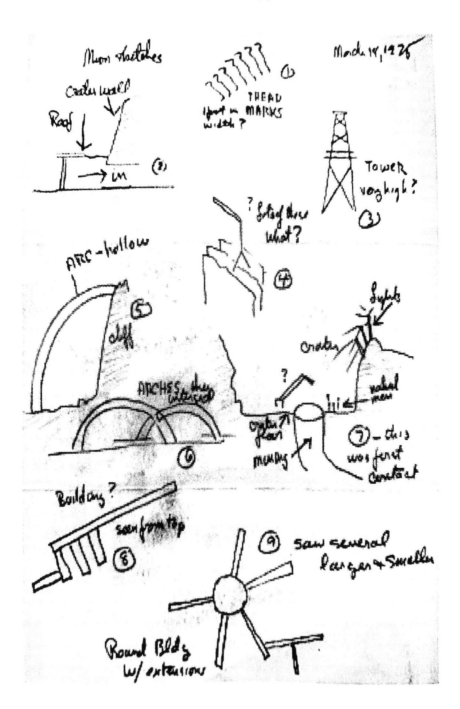

Ingo Swann Moon Sketches
(continued)

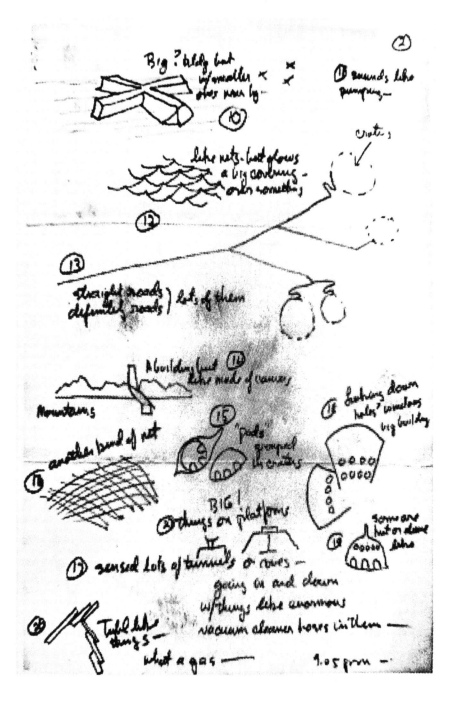

THE EVENT IN LOS ANGELES

During August and September 1976, I traveled several times between SRI and Los Angeles. I went to La-La Land to pursue yet another of the many kinds of studies I had undertaken hopefully to increase my understanding of higher human functions.

I had several good friends in La-La Land, and I was staying in the house of one of them. This was a certain Conrad W. who had many extraordinary qualities. One had the feeling that Conrad was a kind of uncomplicated "old soul," who was somewhat out of place in a modernist society that had become dramatically detached from even remote elements of wisdom.

Conrad seemed to absorb insightful information by telepathic osmosis. He was articulate about many things, but very laid back and with a subtle sense of humor. In other words, he was delightful to be with.

I was also in touch with the marvelous researcher, Dr. Shafica Karagulla, and her research companion, Dr. Viola Neal, both of whom have passed on.

Shafica was a neuropsychiatrist who had broken with mainstream psychiatry to establish the Higher Sense Perception Foundation, and had published in 1967 her famous book, *Breakthrough to Creativity.*

Viola was a notable clairvoyant whose psychic mapping of human biofields and psychic diagnosis of diseases had been confirmed by doctors and in hospitals.

Viola and Shafica were very kind to me in many ways. Their combined knowledge levels seemed extensive, even endless.

One had the impression that their consciousnesses were existing in some higher frequency range—from where they could look down into the dross of average human life.

And in this dross, this dynamic duo could perceive conspiracies everywhere at work—conspiracies erecting invisible prisons into which human abilities were incarcerated and destroyed.

Beyond this, the two were plugged into the unofficial tom-toms of science central and its many fringes which included parapsychology, governmental plans and plots, and the carryings-on of contemporary mystics and occultists.

But they were careful about relaying important gossip because, as they said, doing so was dangerous not only to reputations but in extreme cases to life and limb.

They were interested in my "work," but a little paranoid because of the close governmental connections to the developmental project at Stanford Research Institute. What this meant was that the duo would gossip only to a certain degree about the conspiracies of the world but were careful NOT to say too much.

I, of course, wanted the whole package, for Karagulla was not simply one of those off-the-wall conspiracy buffs dwelling in notable confusions characteristic of at least some conspiracy enthusiasts.

In her earlier days in the Middle East she had actually worked for a number of official intelligence agencies, and in truth she WAS familiar with the internal workings of the vast international sub-rosa organizations.

In order to get deeper into the whole conspiracy package trembling within the duo, I had found that if I plied them with a little good wine, their reticence lost its edge and they talked more freely.

While staying at Conrad's place, I planned another liquid safari into their reticence. Conrad was also a conspiracy addict as was I, and so I asked him if we could have the duo to dinner at his place.

Since gourmet cooking was a passion with me and with Conrad, and since I had explained the full purpose of the meal, we had to plan a superlative menu and lay in supplies for it. The supplies, of course, included reasonably good wines.

Conrad drove us to very large Hollywood supermarket filled with, among other things, the lush produce of the San Fernando Valley farms and orchards.

I had decided to start the meal with artichokes stuffed with crab and breadcrumbs, topped with cheese melted with a fine brandy.

In order to reduce shopping time, I gave Conrad a list of other comestibles. He went in the direction of meats, and I went in the direction of the veggies.

The supermarket had huge tables loaded with artichokes. At one of the artichoke tables was standing a ravishing woman.

She was notable not so much for her excessive female physical endowments, but by the fact that they were barely covered.

She was dressed in the briefest of halters of pink with big yellow polka dots. Beneath that were short-shorts so short they barely existed. Far beneath that she wore a pair of platform high heels about eight inches high.

She had volumes of gorgeous black hair, and her eyes were covered by purple sunglasses. She was absolutely awesome.

I thought: "Good Heavens!!!"

She was rummaging through the artichokes, and I wanted some, too. So I worked my way covertly and nonchalantly into her proximity so I could closer appreciate her near-naked breasts.

To make this look natural, with my eyes barely on them, I fumbled some selected artichokes into a bag.

And then! For absolutely no reason at all I experienced an electrifying wave of goosebumps throughout my whole body.

The hair on my arms practically stood at attention, and the hair on my neck definitely did.

Without rhyme or reason or forethinking or anything at all I suddenly "knew" she was an alien, an extraterrestrial.

My throat went dry. My hands started shaking. And so I decided to back off and start examining the oranges and grapefruits for the fruit aspic Conrad and I were going to try to achieve.

To get to the oranges, I had to turn—and then!

Way down the line-up of vegetable cases I recognized, of ALL astounding and possible things, ONE OF THE TWINS!

HE was watching the woman.

HE saw that I saw him, and there immediately arose in my mind an image of a white card: Please do not speak, and please act normal.

Trying to gather my surprised wits, by now somewhat shattered, the silliest thought then occurred to me.

Well, if one of the twins is HERE, of all places, then the other one must be, too.

And sure enough, the other twin was at the opposite end of the vegetable line-up—and he was watching the woman, too.

This time, both the twins were DRESSED IN BLACK! Not the infamous black suits worn by those who are said to warn those who have seen UFOs not to talk about them.

Here were black jeans, black boots, and black tank-tops, looking like macho hoodlums of the L.A. variety.

At this point, I realized that I was somewhere I should not be, and I made a hasty and strategic retreat to the bread section on the other side of the store. By the time I reached the bread section, a considerable wave of TERROR had begun to make itself felt.

Something here now needs to be clarified.

If I hadn't seen the twins, I might have attributed to my imagination something of the strange effect I experienced near the ultra-sensuous woman.

But that was now not possible at all.

The twins' presence, coupled with my psychic alert, confirmed that the woman WAS an ET.

I don't quite remember how the rest of the shopping went.

Conrad and I got our groceries checked out. On the way to the car, I explained that we would not attempt the orange aspic after all.

Once in his car in the parking lot I asked him to wait a few moments. He asked what was wrong.

I said, "just wait." Shortly, the female came out pushing a loaded grocery cart.

"Study that one and tell me what you think."

Conrad looked briefly at the woman—and then said the most remarkable thing.

"Well, if you mean do I think she's an extraterrestrial, yes," said Conrad in a bored way. "We've got a lot of them here in La-La Land."

I didn't ask him what he thought about the two who were observing her squeezing her fabulousness and groceries into a broken-down yellow Volkswagen. I sank down in Conrad's car and urged him to make a quick departure.

The dinner for Shafica and Viola was a complete success. Naturally, after plying them, and ourselves, with copious amounts of good wine, we told them that we had seen yet another ET down at the supermarket.

This opened up rather inebriated discussions about the ET civilization which was busy infiltrating Earth. Shafica and Viola discussed all of these things in whispers—and the more serious they became the harder it was to hear them.

Viola: "There are a lot of THEM, you know, and many are bio-androids."

Shafica: "They're dangerous, you know, and they realize that Earth psychics are their only enemies. Be careful, Ingo, be careful."

All of THIS information package—without ME ever mentioning the Axelrod event.

GRAND CENTRAL STATION

t was only a few days after The Encounter that I returned to New York for much needed relief from research. I only half expected a call from Mr. Axelrod. It was not long in coming.

The phone rang early one evening, and a cheery female voice on the other end asked: "Mr. Swann?"

I said, "Yes."

"A friend of yours would like to talk with you."

"OK."

"He wants to talk to you on another telephone. Is it convenient for you to be in Grand Central Terminal at 7:30 tonight?"

"I suppose so," I replied.

"Very good, then. Go to the vicinity of the information box in the central concourse and wait there until you see someone you will recognize."

My telephone then abruptly went dead! No good-bye or thank you, no sizzle, static, or dial tone—as if the telephone was out of order.

I picked up the receiver again in a few moments: it was still dead.

I took the subway to Grand Central, and joined the masses thronging around the information box in the grand and very large main hall.

There is a large clock on top of this information box, and I saw I was five minutes early. Those five minutes passed, and so did ten minutes more.

I said to hell with it and went to get a take-out cup of coffee in one of the arcades just off of the main hall. I lit a cigar (these were the days before no-smoking in public places.)

Then, standing about ten feet away, I suddenly saw someone I recognized. I think I had noticed him before, but the fact had not registered.

He, of course, was one of the twins, but dressed in a fashion which made him look like one of the homeless vagrants that hang out in the great railway station.

Seeing that I now recognized him, he put a finger to his lips, and I gathered I was not to show any signs of recognizing him. I don't know why my hands were shaking a little, but they were. I sipped my coffee. The twin spent about ten minutes carefully surveying the floods of people in the terminus.

Finally he gave me a slight nod, and headed in an easterly direction of the arcade away from the great hall. I gathered I was to follow.

He went down one of the corridors leading to Lexington Avenue.

There were, and still are, some stairs leading down to a subway entrance inside this corridor.

Making sure I was in tow, down them he went. I followed.

I next caught sight of him standing near to a bank of telephone booths (which don't exist today.) He stepped into one and I could see him through the glass dialing a number (these were the days when telephone booths with doors still existed).

I kept at a distance, but I am sure he never said a word into the receiver.

He then put the receiver down on the small counter inside the booth and moved away.

I gathered I was supposed to go into the booth and pick up the receiver.

At the other end was a little static, and not knowing what else to say I said "Hello."

"Mr. Swann?" It was the same cheery female who had telephoned me at home.

"Yes."

"What is that thing on your right hand?"

"Oh, you mean my tattoo?"

"What is its color?"

"Mostly green." I replied.

"Good, then. Please wait until you are linked."

Linked? What did THAT mean?

After which followed several beeps and noises and different

forms of static.

Finally, Mr. Axelrod came on.

"I'm sorry to have to do it this way," he began, "but we had to get you to a telephone which scrambles our conversation, and where you can be watched." I was about to say hello, but Axelrod's voice became very firm.

"Do not say anything except answers to my questions." So knowing he was going to yell at me because of my inadvertent encounter in La-La Land, I remained quiet as a mouse.

"I may seem a little aggressive," Axel said, "but we would like to know why you were in that Los Angeles supermarket?"

"I was staying with some friends, and we decided to cook dinner. I wanted orange aspic with lamb chops and I wanted to stuff artichokes. We didn't have any."

Silence. Then: "There was no other reason? Had you ever seen that woman or seen her since?"

"No."

Silence. "Why were you looking at her?"

"Well, for chrissakes, she was extremely sexy and nearly falling out of the few clothes she had on. I first saw her from the rear, and just tried to see what the front looked like up close. She was messing with the artichokes."

"You're sure there was no other reason."

"Absolutely."

More silence. Then: "What did you think of her?"

Now it was my turn to be silent for a moment. "Well, I don't know why—but I got the impression she wasn't, well, exactly like us."

"What WAS she like?"

I nearly choked on the word. "Extraterrestrial!"

"What made you think that?"

"I have no idea. It was just an impression. She had some kind of vibes or something. She sent chills up my spine, and I felt the hair on the back of my neck starting to stand up."

"Have you felt you have seen people like her before?"

"If you mean have I seen extraterrestrials before, the answer is no. Strange people, sure, but nothing like I got from her."

'Why did you run away so fast?"

"After I spotted the twins, I realized something was going on. The whole thing scared the shit out of me."

"OK," Axel said after a pause, "I'll buy it. Do you think she noticed you psyched her out?"

"I have no idea. She was into the artichokes. The whole thing happened too fast. But she never looked at me, at least I couldn't say for sure, since her eyes were hidden behind those strange purple glasses."

"Think, man!" Axel insisted. "This is very important. Did she notice you at all?"

I suddenly started shaking. "No...my best guesstimate is that she did not."

"Were you at the counter first, or was she?"

"Well, she was. I first saw her from way down the aisle, and then I decided to go up and have a closer look."

"You're sure?"

"Sure of what?"

"Well, did she make any attempt to get close to you, or was it you who made the attempt?"

I wanted to blurt out that the twins well knew the answer to this, since apparently, THEY had her under surveillance.

"I don't think she saw me at all, and she was there when I got there!" A tone of desperation had entered into my voice.

Silence. "Good. OK. I feel obliged to tell you that she is very dangerous. If you ever see her again, especially if she approaches you, make every effort to put distance between you and her. But act natural, always do it naturally."

I had no idea what to say, so I said nothing.

"Do you understand?"

"Not really," I managed to whisper, "but I guess so."

"Good. How is your remote-viewing work coming along at SRI?"

By this time sweat was pouring down my sides from my armpits. I was relieved to have the topic change. "Very good. We are getting good results, and I am understanding more every day. I am aiming at least for 65 per cent accuracy across the boards."

"Hum," Axel breathed. "Can you actually achieve that?"

"Probably, but frankly not in every case. We all, our clients too, are interested in the possibility."

Another long silence. "We would be interested in...we have a special task...can you let us know when you reach 65 per cent? How long do you think it will take?"

"Well, if we don't do it soon, we might not get more funds for next year."

More silence, this time a long one. My hand on the receiver was sweating. Finally: "You have an office with a desk in it, right?"

"Yes."

"When you reach 65 per cent, take an ordinary piece of bond paper, 8" by 11", write 65 on it, and leave it beneath the blotter."

How did he know my desk at SRI had a blotter? "OK," I said.

"Good. We will be in touch shortly after that. Do you understand everything?"

Hardly. I understood nothing. But I said, as conspiratorially as possible: "Yes."

"I'm sure," Axel continued, "you get the general drift of all this...that no one, NO ONE should learn of any of this?"

"I get the general drift. All this is serious—and 'dangerous,' right?"

I didn't think I should tell him of our dinner conversation back in Los Angeles where apparently everyone viewed sexy extraterrestrials every day.

"You got it."

Axelrod hung up. These people, whomever they were, never said good-bye or thank you.

The line went dead for a moment as the "link"—whatever it was composed of—was disconnected. But then at least a dial tone came back on.

The twin had apparently seen me hang up, and when I came out of the booth he nonchalantly walked by with a paper cup as if he was soliciting a hand-out.

Attached to the cup was a small card: "Go directly to Lexington and grab a cab. We will guard your rear. Do not look back."

Nervous as hell, but thinking it would appear proper, I boldly pulled out a quarter and plopped it into his cup, where it clanged with other coins.

I went to Lexington Avenue and flagged down a cab as fast as possible, never once looking back.

But I didn't take it directly to my address, but to the corner of Eighth Street and Third Avenue, where I loitered for some time, trying to see if I had been followed.

I then went to my favorite bar nearby and over-consumed cheap beer. My imagination was going full steam ahead.

The paranoid fears that followed this event occupied me for some time thereafter. I had the distinct feeling that everywhere must be extraterrestrials and/or Axel's henchmen or operatives.

And? WHO WERE this Axelrod and his henchmen, anyway?

I spent days, weeks, cycling through the possibilities. CIA, KGB, Mossad, M-5, some ultra-secret military goings-on?

Worst of all was the speculation they, themselves, might be extraterrestrial.

Perhaps there was a space opera going on in which two different sets of extraterrestrial troops were fighting some kind of war here on Earth—while both at the same time were somehow ensuring that HUMANS never realize that they, themselves, are psychic.

What a scenario, huh? Talk about being on the outermost fringes! Fringes so outermost that one didn't even know where the fringes were in relationship to anything else.

The worst thing was I could not talk, certainly did not dare to, about any of this to anyone.

I was sure I had gotten into something quite over my head. I worried I might get killed or kidnapped—disappeared—and end up as slave labor in the mines on the Moon.

Even as I now write, which I am sure many will find too incredible, for words, I have to wonder...

About a year later, in June 1977, I placed the 65 percent signal under the blotter on my desk at SRI in California, which is to say, in our allegedly secure, guarded premises there. Entrance to my office was code locked. Only I knew the code, and it only existed in my mind.

I checked under the blotter every day and afternoon thereafter for about three months.

Then one morning when I lifted the blotter the hair on my arms once again stood up. The signal was gone.

In its place was some dust-like powder in which a finger had scrawled two words: "Expect contact."

I brushed the powder into my trash, and sat down, completely unnerved.

My next meeting with Axel and his crew completely blew my mind, knocked my socks off!

The result of the promised "contact" was that if I had any remaining doubts about whether they existed, such doubts were shortly to be resolved.

I almost got killed in the process.

MR. AXELROD & HIS TRAVELING PLANS

T he expected contact came early in July 1977, a few days after I discovered the message in the dust. The "campus" of Stanford Research Institute has a very nice dining room where my colleagues and I often ate lunch, especially if we were being visited by "dignitaries."

Access to the dining room is through a large lobby, at one end of which stands a very large globe of Earth about six feet in diameter.

I don't remember who we were lunching with that Friday, but when we passed through the crowded lobby on the way to the dining hall, there was Mr. Axelrod standing as large as life, but I suppose inconspicuously, by the globe of Earth.

When he saw that I had noticed him (I had actually stopped dead in my tracks) he walked quickly into the men's room adjoining the lobby.

So I did what I thought he expected of me. I excused myself from my colleagues saying I had to take a leak. To do this, I had to get the key to the toilet from the dining room hostess. Everything at SRI was kept locked because of its Pentagon connections and the fear that terrorists might plant a bomb in the heads.

When I entered the men's room, Axel boldly locked the door with a key and whispered in my ear: "Can you get away right now for the weekend? I want to take you somewhere and show you something? Just nod yes or no."

I nodded yes.

"There's a car in the parking area outside the lobby. I'll wait for you there. Invent a convincing story for your friends. You may be away as long as four days." He then unlocked the door.

I had to think fast about what might stand up as a "convincing story" as I joined our group in the dining room. But all I could really think about was how Axelrod had secured the key to the men's room.

I told my colleagues that I had just remembered I was supposed to join some friends in San Francisco for a long weekend, and without further ado just left them.

The "car" outside proved to be a high-wheeled Jeep, and Axel himself was the driver.

We sped out of the SRI grounds in silence. Axel made it to the freeway heading toward San Jose. Then: "Have you ever seen a UFO?" he asked.

"Yes, I think so."

"Can you describe it?"

"Well, when I was in high school in Tooele, Utah, I used to climb to the top of a large hill called Little Mountain.

"From there you can see across the vast Bonneville Valley and see the Great Salt Lake to the north. It has big islands in it, you know. The view of this vista was just wonderful.

"I used to take naps up there in the late afternoon, but on this particular day I noticed a speck of light really high in the sky over what must have been Salt Lake City.

"It was flying west, and I thought it was an airplane moving really fast.

"But at a certain point in its westward flight it abruptly made a right angle turn downward, not a curved turn down but exactly 90 degrees.

"It plunged straight down and fell into the shadows of the islands or mountains because the sun was lowering in the west and making shadows go to the east.

"I stood up, thinking that the plane had exploded or crashed.

"But as I did the thing rose directly straight up, out of the shadows.

"It rose up to its former elevation, like about 35,000 or 40,000 feet up, and once there disappeared directly into the west in a burst of speed, which was dazzling.

"I didn't know what to think of this, but years later decided it must have been a UFO after I had learned that some of them make right-angle turns.

"Why it did what it did is beyond me. The whole of this down and up and speeding away took place in less than a minute. All I really saw, though, was a speck of light."

Axelrod was silent. It was hot, the Jeep had no air conditioner. Then: "We may have an opportunity to see one of them rather close up. Are you game?"

Even among all the astonishments available in the Axelrod scenario, nothing could have amazed me more. "You mean there's one around here! You've captured one?"

"Oh, no, not that. We have to take a trip, then a hike to a place where one shows up at intervals. Are you game?"

Was I GAME! Who wouldn't be?

Axelrod drove us to the San Jose airport and, leaving the Jeep in a "no-parking" area in front of one of the terminals, we walked directly through the lobby and out the other side to a waiting Learjet.

Other people had flown me in similar planes, wealthy people who were interested in talking about using Psi to discover sunken treasures and oil deposits. I loved the elegant jets for their sense of opulence and power, one of the most poignant status trophies for having "arrived" financially.

Waiting alongside the jet was one of the ubiquitous twins, this time dressed in an olive-green jumpsuit and helmet— definitely "military" in appearance.

We were aloft in about three minutes. It turned out that the other twin was doing the flying.

Once aloft, the twin provided some sandwiches, and Axel said: "We are going to the middle of nowhere. It will be rather cold and rough there.

"But we have all you will need, including a supply of your cigars (he smiled), so after you've eaten you should catch some sleep. It's about a five-hour or so flight.

"It will be dark when we arrive, and then we have to drive about two hours after that.

"Don't ask where we are going, since I can't tell you that—and (with some hesitation) it's better that you not know."

"You know, Axel," I replied, "it may be that I could function better if I knew what was going on."

Axel frowned through the sandwich he was eating. "Well, I can't tell you very much since doing so would endanger our mission and perhaps you, too.

"But I could ask you what YOU think is going on."

So here it was again, the one-way conversation typical of all "Mr. Axelrod" encounters.

"Well, I guess you guys, whoever you are, have a problem, and from all I can tell, Earth is under some kind of siege. UFOs appear everywhere, are seen by thousands.

"Yet they are elusive, but of concern, and so you are trying to fit the pieces together. And I would suppose, too, that you are desperate, enough at least to try to employ psychics to help you out."

"You see," Axel laughed. "I don't have to tell you anything, do I?"

It was no use. So I settled down and tried to sleep, which I actually did even though I thought I would not.

Axel woke me later. "Fasten your seat belt; we're going to land in a few moments." I glanced out the windows. It was dark outside, and there were no lights anywhere.

But shortly we bumped along a runway without the aid of ANY lights.

"No lights?" I commented.

"It's a very high-tech plane," Axel commented. "It just LOOKS like a standard Lear."

Once landed, we descended from the jet into not just cool but icy air, rich with the smell of pines.

Our only lights were flashlights the twins were carrying.

Nearby was a van of some kind, painted with camouflage colors. Nearby I could make out a small building, but it was empty, or at least had no lights in it.

Inside the van: "Here is a jumpsuit," Axel said. "It is thermalized, yet light. You have to remove everything, and you can have no metals on you.

"I know you have teeth fillings, but there is nothing we can do about that. All the fasteners on the jumpsuit and its hood and attached gloves are made of wood and leather."

So shortly I was geared up, finding there were pockets large enough to hold a cigar supply.

While I was changing, the twins had started up the van and we were on our way to wherever we were going.

The drive lasted about two hours. We climbed up some mountains and negotiated some steep hair-pin curves. No one talked at all.

Against the dark sky I could see tall pines whose own darkness blotted out the amazingly beautiful spectacle of billions of stars. We were in the far north, I concluded.

At a certain point, the sound of the van's motor ceased. Yet the van kept moving. I had, and still don't have, no idea of how a van can move without its engine going.

Finally, the van parked beneath some pines.

We disembarked. "We have to walk about forty minutes, now," Axel whispered.

"It's extremely important that we be as quiet as possible. Do exactly as we do, make no noise, and DO NOT talk! And DO NOT light a cigar!"

So we walked in what amounted to almost pure blackness, but at a good pace.

At certain points, one of the twins would take my arm to help me, for instance, across a small creek or around an unseen rock.

They had some kind of goggles over their eyes, which I took to be night-vision things. I didn't at all see why they had not provided me with them, too.

We went up and over some ridges, and then down into some kind of large, flattish area dense with pines. Then we climbed into a sort of arroyo. Once there, we walked a few feet further and sat down on a thick cushion of fallen pine needles behind some large rocks.

Axel whispered: "We're here. Out there in front of us is a small lake. As dawn comes, you'll be able to see it through the pines. We now wait, and hope we are lucky. Say nothing, do NOT make any noise."

"Lucky?" What did that mean?

SEEING ONE

I couldn't see anything at all, save the narrowest dark blue-green glimmer of dawn in the east.

I whispered back to Axelrod: "What am I supposed to do?"

"Just observe, we'll debrief later," he responded. "But it's really important now to observe complete silence from this point on. And do not move unless I tell you to. They detect heat, noise, motion like mad."

So, I was silent.

There we were, four of us sitting silently like rocks ourselves. But suddenly, the two twins gave some kind of hand signal.

"It's begun," Axel whispered. "Please, please! DO NOT make any noise, and do not move unless we tell you to."

My eyeballs rolled around trying to perceive what had begun.

I couldn't see anything unusual at all, save for what appeared to be some gray fog forming up in the direction of the lake. I thought it was just morning fog coming up.

This fog continued forming for about five minutes, and suddenly I saw what had "begun."

For in a moment's eye flicker the gray fog changed, first into luminous neon blue, and then into angry purple.

At that point, Axel and one of the twins put a firm hand on each of my shoulders, and it was a good thing they did.

A network of purple, red, and yellow lightning bolts shot in all crazy directions through the "cloud," and I would have jumped up if not held down.

And then, there it was. Somewhat transparent at first, but in the next second, as if fading-up (like the movie term) out of nowhere, there IT WAS!—solidly visible over the lake whose reflecting waters I could now clearly see.

And IT was GETTING BIGGER!

I don't really know what I had expected, but I had assumed that what I would see, if anything, would be something like a flying saucer. No chance of a saucer here, baby. Because IT was triangular, and its top angle sort of inverted in pulses, so that overall it appeared to be diamond shaped.

At that moment in my astonishment, we could hear a "wind" coming, and it moved past us like a tangible magnetic field, rustling the pine trees around us so much that some cones and branches fell on us.

The two firm hands on my shoulders tightened, warning me not to move in pure physical reaction.

At the same time, ruby-red laser-like beams began shooting out from the "thing," which incredibly was now growing even MORE in size—while still stationary in its original position over the lake.

One of the twins now TALKED softly, although the sound of his voice was like thunder to me.

"Shit! They're enveloping the area! They're going to spot us!"

I had no time to wonder about what he meant. Indeed, some of the laser-red beams had begun blasting pine trees! Of all things!

At the same time, the "thing" had now increased its size to what may have been about ninety feet wide.

The whole of this so far had been accomplished in COMPLETE silence, and even the electric bolts had not "crackled." The blasting of the trees, though, was now audible, while at the same time I could begin to hear low-frequency pulsations.

"They're blasting deer or porcupines or something in the forest," Axel explained softly in a calm but tense stage whisper. "The beams sense biological body heat, and they're sure to hone in on us."

At that moment, the two hands tightened on my shoulders and I was dragged and practically thrown back down into the arroyo.

There was a terrific "pop" where we had been, and some large branches of nearby pines cascaded down on us.

That was my last sight of the triangular thing, but in that last moment I could see the WATER OF THE LAKE SURGING UPWARD— like a waterfall going upward, as if being sucked into the "machine!"

I had landed rather hard on my butt. But with my feet dragging, the twins pulled me up and ran with me between them down the arroyo a short way where they suddenly flung me like a sack of corn under a rock overhang of some kind.

Axel plopped in virtually on top of me, and the four of us huddled packed together like mice in a matchbox.

Axel was breathing hard. The twins were breathing hard.

I was barely breathing at all, and it took some while to realize that a rock or stick had cut through the jumpsuit into my leg, and that the wound was bleeding.

However, I needed no whispered instructions to be as quiet and still as possible.

I was virtually petrified with a kind of terror for which there are few if any words to describe. But there was also a kind of thrill. I HAD seen one!

There we remained for a period which could have lasted anywhere between five minutes to five hours for all I knew.

In this timeless zone, I heard one of the twins say, "All Clear Now,"—which seemed absolutely the most ludicrous thing I had ever heard. If ANYTHING was clear, I had not the faintest idea of what it was.

Axel asked if I was hurt. The twins stood up and, of all things, calmly took leaks while surveying the environment. For the first time, I noticed that the sky was sunlit, the pines gloriously dark green, and that birds had been sounding off for quite some time.

I stood shakily up—and threw up the contents of my stomach no less than three times.

Axel fussed with inspecting my leg wound (not very large, but bloody enough) and I started to say something like: "Yeah, yeah" I snorted, "don't tell anyone, huh!"

"No," Axelrod responded. "I wasn't going to say that, but that it's gone now, and everything's okay."

I stared at him incredulously, and just as irrationally said: "Then it's okay to light up a cigar, huh?" which I proceeded to do by pulling out a pack from the jumpsuit leg pocket.

The cigars were smashed, but while sitting on a rock I licked one of them back together and proceeded to smoke.

One of the twins was limping. The other was nonchalantly cleaning his fingernails with a small stick.

As for ME! Cascades of the most forbidding anger were pulsating through my entire body, making my hands shake.

My reality-hopper was in ultimate wreckage. La-La Land, to the max.

Finally, Axel said that the water in the stream was good to drink, and one of the twins jerked his head as if meaning we should depart, which we did as if just coming back from a hiking expedition.

"So," Axelrod asked while walking, "what did you sense?"

I burst out laughing: "You're completely nuts, Axel! I have to be calm, cool, collected, and in good shape to sense anything. But you can bet your ass you've got a real problem, haven't you!"

Then, from a queasy area of sense-making not exactly intellectually conscious: "It was a 'drone' of some kind-unmanned, controlled from somewhere else—wasn't it?"

Axel frowned, looking at the slope of the hill we were descending. "What was it doing here?" he asked tentatively.

"Well, for chrissakes! It was THIRSTY! Taking on water, obviously. Someone, somewhere needs water...so I suppose they just come and get it.

"You don't need to be psychic to see that! Yeah! That's it, supply 'ship' Earth! Let's drive over to Earth, go shopping and pick up what we need, that kind of thing."

More silent walking until we were again driving down the bumpy road in the van which had un-smashed cigars and sandwiches in it.

"You know, Axel," I finally said, "they're really mean to blast away at 'deer and porcupines'. What can possibly be the sense of that. I've read that some landed UFOs incinerate humans. Is that true?"

Without waiting for an answer, which I knew I wouldn't get, I talked back to myself. "I suppose it is. I guess we would have been blasted, too, wouldn't we? You guys seem used to this, do you do it every once in a while?"

When we finally arrived at the airstrip, which I expected to be a secret one, I found it was thickly populated: with a USA-Alaska mail plane; three Caucasian men in plaid coats and cowboy hats lounging on the wood benches near the small hut; a police pick-up truck replete with two big-bellied "sheriffs"; ten women who I supposed were Eskimo.

All these kept their distance from US.

Near the plane was what might be called a La-La Land special: a hot-dog cart with an orange and blue umbrella.

No one was operating the cart, but the twins walked over to it and made themselves some steaming hot-dogs.

"Want one?" Axel asked. Indeed I did, three of them in fact, dripping with ketchup and mustard.

"Do they know what you are?" I asked, nodding to the observers. At this I finally got an answer to a question!

"Well," Axel responded, "they generally have been told we are wealthy environmentalists and bird-watchers who are assessing acid rain damage."

"Such bullshit," I giggled. "They know what happens up there. That's probably how you found out about this Go-To-Earth Shopping Cart."

The twins had started up the jet. As we lifted off, I could see three of the Eskimo women pushing the hot-dog cart to the hut.

About ten minutes later we passed over a tall range of beautiful mountains, then another, and about forty minutes later over a coastline and out over the ocean.

"Alaska, I suppose. That's what the mail plane said," not expecting an answer, murmuring it to amuse myself.

"Any feelings about how the object transports itself?" Axel asked.

I looked at him and burst out laughing. He had to be kidding! The "object," indeed.

"Well, it must be some kind of 'space displacer', but really Axel, I haven't a clue. But I CAN understand why people who see something like this don't believe it—and why people who haven't seen it CAN'T believe it."

Axel was silent, staring out the window.

I went on: "As I remember it, the thing did not 'transport' itself. It GREW in place right where it appeared.

"It was a pyramidal thing, not a saucer. We think of a saucer flying about, and in fact when we think of things in the air we think of flying in the air.

"We do not think of things growing in place in the air."

Axelrod studied me, but I saw he was perspiring. "Are you sick or something," I asked.

"Ah, I think I cracked a rib when we tumbled. Never mind, it's nothing serious. What's your point here?"

"In our research of remote-viewing capacities, we have learned that when the viewers 'see' something they don't understand, they explain it in ways that make sense to them.

"For example, to a viewer who has never seen an actual atomic reactor, what they are sensing can be described as a teapot, both of which are hot and 'cook'.

"We call this 'analytical overlay', meaning the mind-dynamic process of overlaying something unknown or unrecognized or unfamiliar with a mental image which is recognized.

"The psychic subject in remote-viewing a site with an atomic reactor may well overlay the impressions with a 'teapot' or a 'furnace' because these are the memory images which come closest to what is being psychically sensed.

"If you take the time to let the viewer study diagrams of atomic reactors and photographs of them and their surroundings, the next time they encounter one in psychic seeing they are more likely to identify it correctly rather than call it a teapot.

"But in general, people do something like this all the time. When they encounter something they do not understand, they tend to interpret it in ways they do understand, and they arrive at an interpretation which really doesn't have much to do with what was experienced.

"In other words, they process the unknown through what I call their 'reality-hoppers' and come up with something that fits their present realities—but which may not, and probably does not, pertain to the actual reality of what they experienced.

"People fill in the unknown with what fits with THEIR known.

"The proof here exists in the fact that when five people are shown something which is outside their experience, one of them might say that they don't know what it is.

"But the other four might produce four different explanations of what they saw.

"For example, you called that thing an 'object'. But what I saw materialized, grew in place, and I suppose dematerialized after we tumbled down the rocks and dirt.

"It may have achieved an object status at one point, but to my way of thinking this was an 'appearance' rather than an object.

"A shifting-appearance at that.

"A full part of the problem is that it is a REALITY problem FOR US.

"That thing is outside my reality-experience, and so if you keep asking what I sensed, I am quite likely to begin seizing on overlays to explain it to accommodate you.

"For example, I used the phrase 'space displacer', but I really don't know what that would be or consist of."

Axel twisted in his seat to get more comfortable. "In other words," he commented, "you assess what you experience only within the terms of what you already have experienced, is that it?"

"Pretty much so. Certainly so in experimental tests of remote-viewing, clairvoyance, and sometimes even telepathy.

"But this is a KNOWN phenomenon understood in psychology for quite some time. The only thing is that this understanding is not generally applied to human understanding on a broad scale.

"If it was broadly applied, then it would have to be admitted that what most people believe would turn out to be little more than reality-hopper 'explanations' of what they hadn't understood in the first place.

"We explain what we do not understand THROUGH whatever we think we DO understand.

"I certainly do not understand what I saw at the lake, and I think it is the better part of valor to admit it."

"OK, OK," Axel grimaced. "I got the point—which is that there are two parts of the Problem. What they really are, and what we are going to use to realize what they are."

"Yeah," I giggled. "A remote-viewing novice can study a book with diagrams of all known atomic reactors. You got a book that diagrams all seen UFOs?

"If you had even told me it would be a materializing, levitating triangle, I might not have been so shocked and could have watched it more closely without having my reality-hopper get scrambled."

Axelrod laughed, and changed the subject. "Well, I got your point. It was probably a dangerous risk to expose you to this, ah, appearance, and we really had no right to do it."

I laughed, and relaxed. "Jesus, Axel, I'm ready to go for it again! Who wouldn't be."

"Well, probably that will not be possible. I shouldn't tell you, but our mission will be disbanded shortly, and the work picked up by others, because of strategic security reasons involved."

"Others who will not mix in with psychics, I take it," I giggled.

"You got it. Next week you will be summoned for a complete physical examination, ostensibly in line with overseeing the health status of the people on your project.

"We just want to be sure you experienced no physical damage. The physicians performing the examination will be ordinary doctors who have no knowledge of our existence. Can you explain your leg injury in some sensible way?"

"I won't have time next week. We're going to Catalina Island to do an underwater remote-viewing experiment with a submarine. I'm OK, and the cut on my leg is small. I won't have to explain it to anyone."

The last I saw of Mr. Axelrod was at the San Jose airport, and so there ends the tale of my encounters with him and his ultra-subterranean covert mission.

And I cannot prove a word of that tale and because this WAS the case, I never intended to make a written record of it.

UFOS EVERYWHERE— DENIALS EVERYWHERE, TOO

A s the years passed, two later developments changed my mind about making a written record of what I could remember of the Axelrod Affair.

Both developments somewhat jolted me, but also inspired a certain astonishment about how EASY it had been to forget about that affair.

Indeed, one would think that the Axelrod sequences would have been indelibly etched in one's memory.

But this was not the case at all. And because this was not the case, I slowly became aware that there was something associated with the sequences—something perhaps best described as a kind of amnesia.

The first jolt of memory, however, came about as follows. I am a subscriber to *FATE* magazine, which for a long time was the only publication in the United States that reported on phenomena the modern mainstreams denied could exist.

In late January, 1991, I came across an article by a certain Felix A. Bach, whom I had never heard of.

His article was entitled "Can Moon Illusions MOVE?" The article was headed by a blurb:

"For many years one of our readers has been contributing unique illustrations to objects on the Moon's surface. Now he reveals how he is able to see them—and how you can see them,

According to the article, all one needs is a reasonably powerful telescope. Mr. Bach recommended one with high-resolution powers up to 500x, which the manufacture of Celestron telescopes says allows viewing of lunar details as small across as a football stadium.

Mr. Bach indicated that his telescope was a new SPC 8-inch one set at 600x, and which enabled him to see "at one whack" many towers, nets which come and go, mining equipment, "beads" and "wires," and whole arches.

Bach noted that the structures "come and go" for reasons unapparent and inexplicable, but that they can be seen by telescope is now beyond question.

The article contained sketches made by Bach, some of which more or less resembled the sketches George Leonard placed in his 1976 book *Somebody Else Is On The Moon*. And so, of course, Bach sketches somewhat resembled my own made in 1975 for the benefit of Mr. Axelrod.

Not long after Bach's article, a Ufologist friend of mine visited me, and brought along a book published by some Moon enthusiast in Japan, and which, of course, was all in Japanese.

But the book was about structures on the Moon seen through a telescope. The structures were very hard to identify with the naked eye, but the book provided sketches of what the photos contained. With this help, it became credible that there were structures on the moon.

I eventually tracked down the telephone number of Felix Bach and had several conversations with him. In short, with a small telescope, it takes a "very experienced eye" to separate the vague structures from the lunar background.

This development caused me to consider TELESCOPES, and the details that could be acquired by them. The matter of telescopes will be discussed in the following section.

As a result of the Bach 1991 article, memories of the Axelrod affair now welled up. They had all along been lodged somewhere in memory, but I was somewhat astonished that it took Bach's article to bring them to the surface.

The Axelrod affair had jolted me in more ways than one. Why, then, was it not easy to remember? One would have thought that the memories would constantly have been trembling on the edge of immediate recall.

The second development that helped resurface the Axelrod memories was the advent of Camcorders available to the public.

Before easy-to-purchase video equipment became available, there had, of course, been many still photos of UFO crafts. It was easy to discredit the authenticity of the still photos simply by suggesting that they had been altered in well-equipped photo labs.

This suggestion was cast in cement when SOME photos were proven to be faked, which spread the assumption that all photos "must have been produced" likewise.

Thousands of pieces of independently achieved video footage can hardly be said to have been faked. It might be possible to do so, but only with the help of rather significant amounts of money and extensive digital computer techniques.

During 1991, Earthsiders worldwide had begun to accumulate miles and miles of Camcorder footage of UFOs, many of the craft being simultaneously taped by several recorders, but from different angles and often miles apart.

As a result, a lot of the video footage was shown on various TV shows specializing in oddities—such as *Current Affair,* etc.

While the increasing availability of live video footage is gross in size and dimensions and unavoidable implications, the net result is that official denial did not, and still has not, changed. The sustained denial involves government, scientific, and military attitudes.

That the denial IS SUSTAINED is a patent fact. The question is WHY?

It is 1998 as I write this. I have just read my installment of the weekly *UFO UPDATE* available on the Internet. There are UFOs being rather convincingly reported everywhere. And everywhere, at least in mainstream echelons, the reports are either denied or completely ignored.

Back in 1991, largely because of the Felix Bach article, and largely because I had become aware of memory problems regarding the Axelrod affair, I decided to remember all I could of it and write it down—and to do so before I really did forget it altogether.

You see, I had begun to suspect that Earthsiders as a whole seem to be caught up in some kind of strange but broadly shared amnesia induced, perhaps, in some kind of wholesale way by means totally unrecognizable by human intellects.

I seem to recall, but can't remember where, a science fiction story of social-wide amnesia having to do with hypnotic commands to FORGET, FORGET what you have seen, and ATTACK AND DESTROY those who insist they have seen it.

This kind of thing is really far out, and this writer, of course, cannot insist on anything of the kind.

In Part Three, though, I'll begin to build upon this possibility a little, but only for what it might be worth to the reader.

In order to get at these suggestive issues, we can ignore what has been presented in this Part One, which, after all, is only a personal tale. Instead, we can profit by approaching the issues by reviewing two categories of evidence, even if some of it is only circumstantial.

We will begin by examining some of the evidence about the anomalous nature of the Moon.

MOON
ACTIVITIES

THE MOON AS A TARGET FOR SPIN DOCTORING?

The major purpose of this section is not to present evidence of anything, but to show that evidence does NOT seem to matter.

The answer to WHY it doesn't matter is embedded in an on-going, rather deep and tangled confusion. And it will take mental equipment far more stalwart than mine to penetrate it.

A considerable volume of very strange evidence exists about the Moon that is simply not challengeable even if it is strange.

But since evidence doesn't matter, we need not examine the entirety of the volume to make the point that it doesn't matter.

The above sentence probably sounds like a babble of words. But the babble might become more understandable if the concept of "information management" is factored into it.

Information management has to do with establishing whatever is to be conceptualized as constituting reality within any given societal framework, or within any social grouping.

This is to say that reality-making and information management are somehow interrelated—largely because the reality cannot be constructed unless information relevant to it is managed in this or that way.

The process of constructing a reality requires three activities, all of which require information-management talents: (1) facts, evidence, and information that might support the reality (thereby proving it real) need to be emphasized and, best of all, taught as the truth; (2) facts, evidence and information that would tend to deconstruct the reality need to be disposed of somehow; and, (3) the willful introduction of useful illusions if and when (1) and (2) above cannot be made to seem consistent or to be creatively managed.

Anyone who can become expert in dealing with the three factors above will find his or her talents as a spin doctor much in demand. The term "spin doctor" is of rather recent vintage, of course, but the craft it represents is a quite old and traditional one.

The factors briefly outlined above seem straightforward enough, especially in that the activities of spin doctors are sometimes mentioned in mainstream media.

Indeed, what DOES a spin doctor DO if not to manage information along the lines of the three factors outlined?

Spin doctoring is usually mentioned only in relationship to political intrigues, most specifically those emerging or on-going at the highest echelons in Washington, D.C.

But the extended fact of the matter is that science, philosophy, economics, and sociology are also cluttered with spin doctoring or residues of them.

It is necessary to dig a little deeper in order to achieve a more firm grasp of the meanings of the above considerations.

It can be observed, at least hypothetically, that the average human being cannot function very well unless a few realities are established, and which thereafter arouse some semblance of certainty.

Otherwise, the human sort of flops about in a sinkhole of uncertainty. And groups of humans flopping about in this way are known to be detrimental with regard to any number of situations.

Indeed, information NEEDS to be managed in order to avoid extensive demonstrations of such flopping around.

This natural and perpetual NEED automatically calls forth the profession of spin doctoring.

IF human specimens were NOT information processing entities each, in their own right, then the age-old profession of spin doctoring would not be needed.

As it is, though, human specimens ARE information processing entities. That this is so results in a variety of problems that sometimes stress the creative energies of spin doctors in no small way.

But regarding this there are sometimes two blessings that ease the job of proficient spin doctors.

The first blessing is that a fair share of humans don't really require copious amounts of information as they proceed from the womb of birth to the maw of death.

The second blessing is that a fair share of humans will accept illusory information since it does take rather copious amounts of information to distinguish illusory from other kinds of information.

Spin doctors since antiquity have apparently been keenly aware of these two blessings. It has long been understood that illusory information can serve quite excellent purposes, since in large part there is no professed desire among the human masses for other kinds of information.

To this must be added one other significant factor that not only eases the job of spin doctors but elevates it into a kind of power threshold.

In that each human specimen is an information processing entity, each is also an information processing mechanism.

The latter here is not mentioned in order to diminish in any way the former.

The point is that information is composed of very numerous data bytes, the whole of which needs to be organized into some kind of information whole.

So one fact in itself tends to have little meaning unless it is compared to and combined with lots of other facts. In this sense, then, one fact (or even ten of them) only adds up to a very tiny information universe.

Beyond this, it is quite well understood that the human species, universally and generically, shares in mental equipment which processes information in what amounts to mechanistic ways.

Perhaps this can better be understood by picturing the generic mental equipment as the computer hardware, and then picturing the information (real, artificial, or illusory) entered into it as mental software programs.

This is to say (as those who are computer-efficient realize) that information that conflicts is mechanically rejected, while information that agrees is mechanically accepted.

All the complexities of human mental apparatus considered, this analogy may be a weak one, to be sure. But this situation is, of course, an additional blessing for information spinning experts, or at least for proficient ones.

All that really needs to be done in order to maintain spin control of mental software programs is: on the one hand, to enter and maintain certain data-information factoids considered desirable; AND on the other hand, to delete and maintain as deleted certain other data-information factoids considered undesirable.

Lastly, all human entities seem to possess what might be referred to as an "information comfort zone." Furthermore, most humans seem to LIKE this zone—and they don't particularly want information discomfort to be induced or intruded into it.

Thus, old and familiar information associated with information comfort is desirable over new, unfamiliar and alien information that induces discomfort.

Thus, it is easy enough to see WHY information that might entirely wreck information comfort zones is viewed with ultimate distaste.

After all of the foregoing considerations, it is easy to suggest that the appearance of possible ET factors are factors that can be the source of major wreckage within information comfort zones typical of Earthside information packages.

That Earthside realities ARE constructed is absolutely known because an attentive historical study in this regard clearly shows that different peoples, times, and places have utilized different reality constructs.

The consideration central to all of the above has to do with WHO constructs the realities. Earthsiders, of course, assume that Earthsiders themselves construct their realities.

But if ENOUGH information based in identified and proven factoids is assembled and aligned, a somewhat unnerving answer becomes at least partially visible: WHO it is that constructs the realities is NOT at all clear.

The whole of this, of course, ends up as a mental and emotive quagmire that stresses a lot of brains and synapses.

But it can be noted that the stresses probably come about because most people on their own determination do not construct realities, but more or less adapt to those around them.

Indeed, most people are culturally discouraged from making their own reality treks into reality constructing.

This of course makes possible that great and time-honored profession referred to as reality management by the few on behalf of what the few view as the somewhat dumbed-down majority.

Of all possible candidates that might require the services of reality managing, the Moon seems the least likely.

But if enough Moon facts and factoids are assembled and aligned, then it turns out that the Moon has been treated to exceedingly expansive doses of spin doctor machinations.

There has to be a very good reason as to WHY the Moon, of all things, needs extensive spin doctor treatment.

THE MOON- EARTH'S NATURAL SATELLITE

B riefly summarized, the confused tangle regarding the Moon consists of five major aspects or situations:

1. The traditional mainstream description of the Moon as a dead, airless, natural satellite of Earth, formed at the same time Earth was.

2. Modern scientific technology has revealed that the Moon is nothing of the kind.

3. The modern mainstreams (scientific, military, political, cultural) continue to insist that the Moon is a dead, natural satellite of Earth.

4. The dimensions of the dissembling propaganda needed to maintain (3) in the face of (2) is huge, so huge as to be unbelievable.

5. For (3) to continue to exist, and for (4) to be implemented and enforced, something VERY BIG must indeed be at stake.

As of 1975 and my encounter with Mr. Axelrod, I did not know much more about the Moon than any other average person. However, in the years that followed I took a deep interest in it (for reasons now somewhat obvious).

Others did also, especially when more and more information began to be scientifically discovered and confirmed about the satellite's utter radical nature.

This radical nature is sufficient enough to seriously crack not only the eggshells of conventional Earthside knowledge of the Moon, but many other Spaceside factors as well.

By reviewing but a few of the radical Moon factors, we will encounter a rather sordid mystery, some few elements of which might not be noticed unless they are pointed up.

A first element is that the strange Moon factors that have been discovered to date are NOT exactly covered up, since a good portion of them have been revealed in published scientific papers.

A second element involves not the radical factors which have seen the light of day in print, but rather their direct implications. In this sense it is quite clear that the direct implications ARE covered up.

The exact nature of this cover-up is entirely difficult to pin down. But whatever its intimate details, all such cover-ups gain their awesome efficiency because most people are entirely ignorant about what REALLY is involved.

And where widespread ignorance exists, knowledgeable elite can form whose spin doctors can set about erecting webs of disinformation, which are accepted as information by the uninformed.

If it chances that "official secrecy" needs to be factored into the cover-up, then an elite can form within the elite, and so a cover-up can involve the workings of several elitist strata—to the degree that few can grok from where the core of the cover-up is being managed.

It is because of all this entwining of cover-ups, each of which can have different motives and purposes, that difficulties arise regarding penetrating what's involved.

In this case, the general ignorance involves knowledge about moons in general. Few have any interest in them to begin with. Those who do are limited by material considerations involving telescopes, as we will see ahead.

But if one doesn't know something about moons in general, then one will have no basis for recognizing something that might not be a moon to begin with.

The very basic and widely taught mainstream assumptions regarding moons are in four parts: (1) that planets have them, because they do; (2) that the moons are solid natural formations, as are the planets; (3) that the planets and their moons are made of stellar matter, the matter being in the form of compacted elements pulled together into a ball by gravitational forces; and, (4) that the moons are formed when the planets are formed unless the strong gravity of a planet somehow "captures" an asteroid that comes racing by and then goes into a safe orbit around the planet. This possibility, however, would certainly be a delicate business.

The two inferior planets (Mercury and Venus) of our own star system have no moons, Earth has one, and each of the superior planets (Mars, Jupiter, etc.) have one or more.

Since the advent of modern scientific times, the formation of the Earth-Moon system is dated at 4.5 billion years ago. There has never been any doubt about the assumption that Earth's Moon is a natural satellite formed when Earth was formed, and hence formed from the same general materials.

So far, so good, right?

When Earthsiders began dreaming of going to the Moon, it was clear that they would have to do so by being encased in a spaceship. Further dreaming involved the concept of artificial satellites that might orbit planets.

Out of this arose a small necessity to distinguish between a natural and an artificial satellite. This distinction was logical and incorporated the idea that an artificial satellite needed to be hollow in order to be useful—whereas a natural satellite, such as a moon, obviously would be solid.

In this regard, and as began to be noted during the 1960s and 1970s, scientific discoveries about Earth's Moon produced some rather confusing data about its physical characteristics.

As to Moon's physical characteristics prior to the space age, all official sources of information were based on telescopic and photographic studies of its surface.

After the American and Soviet missions to the Moon, the instruments carried by unmanned and manned vehicles made it possible to extend more intimate knowledge about the natural satellite.

As will be discussed later, knowledge about the Moon acquired in THIS way was to prove difficult to grasp.

The Moon probes were of several kinds, the multitudes of them commencing after the Soviet Union established in Earth orbit the artificial satellite Sputnik 1 in October, 1957.

This was a scientific and/or military coup by the Soviet Union, a Cold War one-upsmanship which embarrassed the United States.

The United States made up for this affront when, in 1961, President Kennedy committed the United States to the goal of landing men on the Moon and bringing them safely back. As is often stated, the resulting Apollo program then became the largest scientific and technological undertaking in history.

But the general Cold War idea regarding the Moon was that the first Earthside superpower to colonize it could rule Earth from this Spaceside natural satellite.

That nothing of the kind happened now is a factoid that has to be carried in mind from this point on.

With the goals of lunar conquest and scientific advancement now firmly in hand, the United States began by launching a quite large number of Moon orbiting satellites that preceded the landings on the lunar surface.

Reported numbers of such satellites vary between fifty and 450, but most specify that the majority of them had military purposes, the truth of which can be imagined with some degree of certainty.

As the Earthside nascent space age developed, probes to the Moon were in the form of fly-bys and unmanned landers: the Soviet Luna crafts, the American Pioneer, Ranger, and Surveyor vehicles. The first landers were designed to crash on the Moon, but shortly soft landings succeeded.

In August, 1966, the United States successfully launched the first Lunar Orbiter, which took pictures of both sides of the Moon, as well as the first pictures of Earth from the Moon's vicinity.

The primary mission of the Orbiter crafts was to locate suitable landing sites for Apollo, the American manned spacecraft program.

Between 1966 and 1968, more American Surveyors were launched, as well as more Soviet Luna Orbiters.

The ultimate goal, of course, was to put men on the Moon. Twenty Apollo missions were planned in order to achieve this goal, the first six being unmanned checkouts of the delicate equipment.

The principal objective was finally achieved in July, 1969, when Apollo 11 was the first to land men on the Moon, while in July, 1971, Apollo 15 marked the first use of the Lunar Rover.

In December, 1972, Apollo 17 was the LAST American craft to the Moon. At that point, American Moon visits abruptly ceased for reasons that were never adequately explained. The remaining three Apollo craft which were already built at enormous expense, were left to rot.

It was not until 1995, some twenty-three years later, that the Clementine craft was sent to the Moon. This, however, was a U.S. Army project, not a NASA effort.

Luna 21 of January 1973 seemed to be the last Soviet craft, but in August 1976, Luna 24 landed on the lunar surface—after which the Soviets, too, desisted from tackling the Moon.

Thus, the great and very expensive race of the two Earthside superpowers to colonize the Moon came to an ignominious end—and for reasons which were not clear at all.

All things considered, especially the enormous advantages of Moon colonization, the reasons must have been quite impressive.

Instead, in 1972, a change of emphasis came about regarding Space Exploration.

After a decade of hot and exceedingly competitive and exceedingly expensive interest in the Moon (and which included the idea of setting up Moonbases there), the Americans and the Soviets decided TO JOIN UP and attempt to erect not a Moonbase, but an Earth-orbiting Skylab.

Thereafter, Moon interest faded into official and popular obscurity—even though the Moon itself IS an Earth-orbiting thing. If one thinks about it, this is somewhat strange.

Additionally, most sources, such as the Columbia Encyclopedia, regarding Space Exploration indicate that Moon missions produced "increasingly large amounts of scientific data."

Even so, until 1997, official scientific descriptions of the nature of the Moon remained more or less the same as they had been offered up in 1957 some forty-five years earlier.

The Moon remained a dead satellite, airless, with high

mountains, craters, and dry, dusty, glassy and stony plains (called Mares (Seas)) estimated to be formed from magma, breccias, and glassy melt resulting from molten sprays of superheated meteorite impacts.

The Moon's age was still given as 4.5 billion old, dating from the time that the rest of the solar system was formed.

SITUATION MOON ROCK

There is a rather sprightly category of information about the Moon called anomalies—the common definitions of this term referring to an irregularity or a deviation from the common rule.

However, a more precise definition refers to something that prevailing wisdom is quite certain cannot exist or is impossible—but which is found to be existing anyway and is therefore not impossible.

In this sense, an anomaly is something that is discovered to exist, and which thus tends to disestablish the comfort zones of prevailing wisdom.

It is quite difficult to integrate the discovery of anomalies into the knowledge systems that have vigorously established their impossibility.

This is an especially embarrassing prospect with regard to the sciences which have somewhat of a vested interest in being correct so as to justify the funds pumped into them.

So, as might be anticipated, the scientific solution to anomalies is to cover them up on the one hand, and on the other to prevent their implications from dribbling down into broader interest and appreciation.

The scientific disinterest in anomalies works quite neatly for the purposes, for example, of elitist spin doctors who are interested in secrecy and want to manage certain kinds of information away from public cognizance.

Here, it is useful to emphasize that an anomaly is not a mere speculation, but something that has been shown to exist.

It is easy enough to comprehend that if there are very good reasons to ensure and to continue to ensure that the Moon is a dead, dry, desolate, NATURAL satellite, then the dribbling down of Moon anomaly factoids that more or less suggest otherwise really would need to be covered up.

Thus, one can expect to find at least tacit cooperation between the sciences (which are embarrassed by anomalies) and secretive enclaves (which desire to keep the anomaly implications from public cognizance).

Since the science systems and the secretive systems reinforce each other along these lines, it is difficult to grasp the end of the thread that might help unravel the resulting cover-ups.

It is now helpful to reprise the main structure of the cover-up confusions.

Between 1957 to roughly the present, the official descriptions of the Moon underwent hardly any change. Meanwhile, beginning in 1961, the Moon was subjected to the most extensive and most expensive technological effort in history.

Most official sources published after, say, 1975, indicated that the technological effort provided "increasing amounts of data"— AFTER WHICH, the official descriptions remained roughly the same as in 1957.

Anyone who has an interest in tracking the existence of anomalies will realize that most of them have rather amusing factors that go along with them.

This is certainly the case in the amusing matter of Earth rocks and Moon rocks. It is scientifically accepted today that Earth and accompanying Moon are as old as the solar system, whose age is dated back to 4.5 billion years.

Due to admittedly wonderful scientific advances, the age of rocks can scientifically be dated by examining tracks burned into them by cosmic rays.

By this technique of measuring, the oldest Earth rocks found so far date only to 3.5 billion years ago.

The Moon missions returned some 900 pounds of rocks and soil samples. From these, a curious factoid was ultimately revealed in 1973: some of the Moon rocks dated back to 5.3 billion years ago.

Thus, between Earth and Moon, this factoid leaves an amusing discrepancy of some 2 billion years, with the errant Moon rocks existing some 1 billion or so years before the solar system was formed.

Then there is the matter of the Moon dust in which the errant Moon rocks were found. The dust proved to be a billion years older than the rocks themselves.

If one is INTERESTED in this kind of situation, here is a conundrum of no small magnitude—in that one has to wonder about two possibilities:

Ø where the rocks and dust came from; or,

Ø where the Moon was BEFORE our lovely solar system was formed, and HOW the Moon got to the solar system and into such a comfortable orbit around Earth.

As will be mentioned ahead, in spite of the now abundant data, the fundamental anomalies of Moon geology remain somewhat confusing.

Generally speaking, the Moon has three distinct layers of rock, all three combined reaching down to a depth of 150 miles.

If the Moon and Earth were formed at the same time, then the material composition of the layers of both should somewhat match. However, iron is abundant regarding the Earth, but quite rare on the Moon.

As the writer, Earl Ubell, noted (in *The New York Times Magazine* of April 16, 1972), the differences suggest that Earth and Moon came into being far from each other, and presumably under different formative circumstances.

The important significance of Ubell's published consideration is that the anomalous differences were established and accepted scientifically; otherwise they would not have been published in the venerable newspaper.

This anomaly confuses conventional astrophysicists with regard to explaining exactly how the Moon became a satellite of Earth— and so there has not been much mention of this upsetting factoid since 1972.

The sum of all these factoids seems to add up as: that Moon, and Earth were formed neither at the same time nor in the same place, meaning that the Moon "came" from somewhere else.

Moving on from rocks, the Moon's mean density is 3.34 grams

per cubic centimeter—as contrasted to the Earth's mean density of 5.5 grams per cubic centimeter.

The meaning of this is a little difficult to grasp, so I'll try to simplify. If Moon and Earth were formed at the same time, and of relatively the same materials, then their mean densities should be somewhat similar.

Furthermore, the differences in the mean density imply that the Moon probably has no solid core, as does Earth, and it is the absence of the core which accounts for the density differences.

If this prospect is pursued to its logical conclusion, then the deep interior of the Moon is along the lines of being hollow.

A NATURAL SATELLITE CANNOT BE HOLLOW

The probability that the lunar satellite was NOT solid was first mentioned in 1962, and was of course immediately challenged as being based in "faulty data."

So several new studies were undertaken. But these new studies ended up with much the same result.

Finally, Dr. Sean C. Solomon of MIT reported (in *Astronautics*, February 1962) that "The Lunar Orbiter experiments vastly improved our knowledge of the moon's gravitational field indicating the frightening possibility that the moon might be hollow."

Frightening? What, indeed, is the significance of that word?

The significance was mentioned by no less a figure than the late and great astronomer Carl Sagan in his book *Intelligent Life in the Universe* (1966).

According to Dr. Sagan, who surely would have known what he was talking about, "A natural satellite cannot be a hollow object."

For clarity here, a hollow satellite cannot be a natural satellite. But a hollow satellite could be an artificial satellite. "Artificial" means made or constructed.

It is suitable to recall here that the decision to put a man on the Moon manifested as the 1961 Apollo program.

The decision at that point would have been based on all earlier available information about the Moon.

By 1962-1963, this information would have included the confirmed possibility that the Moon was either hollow, or at least contained significant "negative mascons."

"Negative mascons" translates as large areas inside the Moon where there is either matter much less dense than the rest of the

Moon, or empty cavities much larger and deeper than Earthside Mammoth Caves, etc.

But, squarely stated, this unquestionably means that the Soviets and the Americans fully anticipated arriving at the Moon which was already known to be more or less hollow. By extension, it was therefore KNOWN to be a satellite NOT of natural origin.

It would have also been clear that the two Earthside superpowers fully expected to utilize the lunar cavities as opportunistic Moonbase habitation.

This "plan," however, seems not to have been fulfilled. One is forced to wonder WHY. It certainly seems an easy enough project, all NORMAL things considered.

Additional and more dramatic confirmation, of the hollow Moon possibility came in November, 1969, when the crew of Apollo 12 sent the ascent stage of their lift-off module crashing back to the lunar surface.

The impact caused an artificial Moonquake.

Ultra-sensitive seismic equipment installed on the Moon's surface recorded that the entire Moon reverberated like a bell for nearly an hour.

As one scientist (among a number of others) indicated, he would rather "not make an interpretation right now."

The evaded "interpretation" could only have been that the Moon was ringing like a bell, and that like a bell it was significantly hollow—not merely having a few negative mascons.

Later, other significant experiments were undertaken to determine whether the Moon was hollow or not.

The important aspect of these later experiments is that their results have NOT been made public.

It does not take all that much intuition to conclude that the Moon is hollow, or something along those lines.

What is not really understood, very broadly at least, is that this factoid was officially known at least by the late 1950s.

As Carl Sagan indicated, if a natural Moon satellite cannot be hollow, then the Moon is NOT a natural satellite.

Even though this phenomenon was amazing, it seems that it could have been taken in stride and that the two superpowers would have proceeded to colonize and to inhabit the Moon's cavities.

It is now publicly known that great plans WERE prepared for Moonbases, which included installing Moon-based missile defense/attack systems.

Yet, this great scheme to colonize and inhabit the Moon did NOT take place. Since this kind of thing would have been cheaper than trying to build a Skylab, one is stimulated to wonder why.

We must leave this wonderment open for a while, but only to avoid becoming mired in mere speculation. There are additional anomalies that will help to avoid speculative conundrums. To enter into the additional anomalies, we first have to examine the matter of telescopes and high-resolution photos.

THE "MISSING" HIGH-RESOLUTION EVIDENCE REGARDING THE MOON

The multiple factors of this chapter altogether represent one of those quagmires one can unsuspectingly get sucked into with regard to the topics of this book. So, it is perhaps the better part of valor here to simply state what this chapter expands upon.

High-resolution evidence of the Moon showing small details of its surface can be acquired only by very expensive and sensitive equipment. Because of its costs, none of this equipment is affordable at the public level. The equipment does exist, but it is under official control.

It is abundantly clear that high-resolution evidence about the Moon has been acquired officially, but none of it has been provided to the public.

Instead, the official sources continue to release only low-resolution evidence, none of which shows smaller details of the lunar surface.

The low-resolution evidence is apparently in keeping with the Dead Moon Dictum—while it is almost certain that high-resolution evidence would present an entirely different understanding of the Moon.

The details of all of this are not without their interest, since one of the principal conclusions to be made from them is that the official Earthside cover-up of Spaceside activity (obviously going on) could NOT be maintained if high-resolution evidence of the Moon was released.

To get into the details, the concept that the Moon is a dead, airless satellite has been widely established, but only because all of the AVAILABLE hard evidence is accepted as confirming it.

The hard evidence exists in two forms: what can be seen through telescopes; and what can be identified from photographs acquired by space vehicles orbiting or landing on the lunar surface.

The AVAILABLE low-resolution evidence is so massive and so evidential that altogether it represents a form of logical certitude that gives an unquestioned basis to the accepted logic of the dead Moon idea.

It is always quite difficult to inquire into matters that have achieved strong degrees of logical certitude. There is a particular reason for this, albeit one not usually recognized.

The reason is that anything that conflicts with the established logical certitude is automatically taken to be illogical and is ultimately treated in a "can't be" way. Thus, anything along these lines is fraught with difficulties—because what is at stake is no longer only the evidence per se, but the logical certitude derived from it.

As many sociologists have observed, this is almost the same as saying that conflicting evidence will NOT be admitted as evidence and will be stigmatized as illogical—especially if there is an official advantage in doing so.

It is therefore quite surprising to discover that the overall situation regarding the Moon constitutes a clear-cut example of this kind of quagmire. "Clear cut" because there are TWO kinds of EVIDENCE.

The first can be referred to as low-resolution (logical) evidence, the second as high-resolution (illogical) evidence. The important distinctions between the two are not all that hard to sort out.

Ever since telescopes were developed in the early seventeenth century (during the time of Galileo), there has been much enthusiasm for developing increasingly refined equipment to make visible on the Moon what is invisible to the unaided human eye.

Earthsiders can of course see the Moon with their eyes. And on a clear night Earthside eyes can even vaguely resolve the outlines of some of the larger lunar geography. But one needs telescopes of higher magnification to clearly see more detail—to see smaller and smaller aspects of the lunar surface.

Roughly speaking, higher magnification equates to higher-resolution, and which renders visible smaller and smaller things. And with this in mind, we now come upon the matter of Earthside equipment the capabilities of which MIGHT make visible small details of the lunar surface.

As everyone knows, a lot of photographs of the Moon have come into existence, especially with the advent of the Space Age and the initial Earthside intents to colonize it. Some of the earlier photos were achieved via telescopes, but later ones by cameras aboard lunar craft.

One can examine these numerous photographs and see the Moon—see its "dead" surface all pockmarked with craters and barren deserts called Seas. Many photos show rills, valleys, canyons, mountains, things that stick up and cast long shadows, and large and small "domes" (which sometimes appear and disappear.)

And most people are content in seeing the Moon via such photos, for THERE indeed IS the Moonscape almost exactly as it logically conforms to the Dead Moon Dictum.

If, however, one introduces the question of what can be seen of the Moon via WHAT EQUIPMENT, one slowly but surely will become involved in officially-endorsed information distortions that smell of some kind of fortuitous cover-up.

This odiferous issue revolves around the distinction between high-resolution and low-resolution evidence regarding the Moon—coupled with the fact that all officially available photo-visual evidence of the lunar surface is consistently of rather low-resolution.

That this should NOT be the case is abundantly clear. Prior to the 1950s, most Earthside telescopes had a resolution of only about one to two miles—which meant that something the width of a mile would be seen as something not much larger than a dot.

But during the 1950s and early 1960s a number of scientific and popular science sources referred to higher-resolution telescopes under development—telescopes so sensitive that one would be able to identify a basketball or a dime on the Moon.

There can be no doubt that such telescopes were developed.

To help ensure clarity it is useful to summarize what high-resolution might consist of—either via telescopic or camera equipment. After all, through the years the lunar surface has been viewed via both.

As but one example of HIGH-RESOLUTION, various media sources have referred to the existence of the so-called Spy-In-The-Sky satellites.

These are said to carry different types of "monitoring equipment," some of which are cameras so sensitive and of such high-resolution that they can zoom in and read auto license plates and bubble-gum wrappers in gutters.

To complete the picture here, it is meaningful to consider the elevation from which this high-resolution can be achieved. Earth-orbiting satellites have to be higher than 200 miles, because if not then atmospheric drag will slow them down and cause them to fall back to Earth.

Thus, there are satellites orbiting Earth at elevations between 400 miles to 22,000 miles. The exact orbits of Sky-Spy satellites are closely guarded secrets. But any reading of bubble-gum wrappers from above 200 miles is marvelous, of course, and a definite kudo for science and technology.

The Moon has a mean distance from Earth of about 238,857 miles. Across that distance, lunar features can be somewhat magnified by average telescopes commercially available to amateur observers.

Astronomy magazines and catalogs advertise what kinds of telescopes are commercially available and affordable. These range from 3-inch to 14-inch telescopes, but much will depend on whatever resolution factors are built into them.

Generally speaking, however, only very large lunar features will come into resolution (or into focus), but smaller features will not, all depending on the equipment involved. Something the size of a large baseball stadium may or may not be seen, but if so would look like a dot. Something also depends on Earthside viewing conditions, such as air clarity, lack of smog, clouds or light pollution, and so forth.

Beyond the public capability, however, most of the sources I've consulted indicate that high-resolution telescopes WERE secretly utilized at the beginning of the Space Age in the late 1950s. The telescopes usually referred to are those of the Naval Observatory and Mt. Wilson, and Mt. Palomar.

The large 200-inch reflecting telescope at Hale Observatories at Mt. Palomar became operational in 1948. But prior to that, Mt. Palomar also possessed 40-inch, 100-inch and 150-inch telescopes. The 120-inch telescope at Lick Observatory became operational in 1959. (The largest commercially available telescope is about 8-inches to 14-inches.)

It may be true that the large reflector telescopes might be a bit unwieldy regarding the closeness of the Moon. But the suggestive situation here is that if one was planning to go to the Moon, and if one had these government-funded, large telescopes at hand, would not one try to utilize them to spy on the Moon in order to anticipate what would be encountered there?

It is quite well known that "secret government" work has gone on at both of these and other esteemed observatories and without doubt at others as well. And it requires no great leap to understand that inspection of the Moon would certainly have taken place by any and all means possible.

Thus, there can be no question at all that at least some of the much higher-resolution telescopes were utilized to espy the Moon.

Indeed, if Moon gazing via these larger telescopes was not undertaken, then the failure to do so is so gross as to bring the concept of human intelligence into question. As it has turned out, though, between 1948 and the present no high-resolution information has been released.

With regard to cameras, this of course refers to television apparatus sent to the Moon aboard spacecraft. But once at the Moon, cameras themselves have different resolution capabilities, depending on their focal length and depth and how distant or close they are to the lunar surface—or to what is being photographed.

There are different technical definitions for high and low-resolution. A general understanding, however, is that low-resolution does not make smaller things visible, and that high-resolution does.

Now we come to the quagmire situation that rotates around three factors: (1) high-resolution equipment; (2) the Moon; and (3) the complete absence of high-resolution evidence regarding the lunar surface.

The situation is further complicated by two additional factors. High-resolution equipment is very expensive, and thus is always under some sort of official control. Many private individuals, however, have quite good telescope equipment. Although their equipment is of lower resolution, many of these individuals have none the less identified rather remarkable structures and activity on the Moon.

However, these individuals (world-wide) are outside of the official network, and so their evidence can be disposed of as being illogical or imaginary in nature.

As a result, a quite active sub-culture, replete with its literature and photographs, has emerged involving the latter, especially in Japan, where fascinating books and TV shows have been produced.

Even a cursory examination of the works of this subculture indicate that there is "stuff' on the Moon that really needs higher-resolution backup. But this is the stuff whose existence is officially denied—while at the same time only official sources have the money and power to achieve high-resolution evidence.

A principle question emerges from this turgid situation. Considering the technological advances of the Space Age, SHOULD higher-resolution evidence of the Moon have been acquired?

This question can be reversed with equal efficiency. Has higher-resolution evidence of the lunar surface NOT been acquired?

Whatever the answers might be, one thing is certain. With one exception, hardly anything even near high-resolution evidence of the lunar surface has ever been made available to the public. The exception consists of few early cases where official lunar photographs were released BEFORE officialdom realized they did contain details unfortunate to the dead, uninhabited Moon Dictum.

To be sure, those released photos were NOT of very high-resolution quality, but even so they did reveal "unusual" phenomena to the perceptive eye common to experienced photo analysts.

And one individual who seems to have such an eye was George Leonard (already mentioned in chapter 6), and who produced, in 1976, a hardback book entitled *Somebody Else Is on the Moon: What NASA Knows But Won't Divulge!*

Leonard acquired official NASA photos and submitted them to a refined analysis. What was not being divulged, according to Leonard, was that a "a highly advanced underground civilization" was working the surface of the Moon, and was busily engaged in mining, manufacturing, communicating, and "improving" the lunar surface.

Leonard's book reflected careful reasoning and respectable logic and was substantiated with thirty-five photographs released from NASA. There was a list of them at the beginning of the book, and their NASA photo numbers were identified.

Based upon the photographs, Leonard detected a rather large array of artificial lunar activity going on, and which, if truly existing, could only be attributed to some kind of ET intelligence.

In addition to the photos, the book was supported by a fair amount of NASA and other scientific documents, and by transcripts of conversation between the astronauts and Houston.

Even though there are low-resolution difficulties with Leonard's book, it has quite a bit in common with any number of amateur and professional Moon watchers before and after him (the sub-culture already mentioned), who have fortuitously stumbled across unusual and surprising lunar phenomena and activity.

Although individuals in several countries (Japan, the United States, England, and South Africa) independently have made contributions along these lines, evidence for lunar structures and artificial activity remains hampered by the lack of higher-resolution evidence.

What has been needed all along (and still is), of course, are photos of higher-resolution. Back in the late 1970s, those interested in this matter assumed that such higher-resolution evidence would soon be forthcoming—largely because it was unthinkable that they had NOT been acquired by the many lunar Flybys, Orbiters, Landers, and manned missions, and which altogether returned some 140,000 lunar photos.

This evidence is nowhere forthcoming. Instead, the public has consistently been provided only with photos of low resolution. Over time, such photos have been released several times to different individuals.

When later versions were compared with the earlier same versions, certain photographed areas showed dissimilarities—as to suggest evidence of retouching and the elimination of factors too revealing. Also, many official photos seem rather conveniently to have been cropped just shy of suggestive areas.

Additionally, Orbiter photos of certain areas have never been released—such as those areas around Mare Crisium and the craters of Plato and Aristarchus where activity suggestive of "inhabitation" is known to take place.

All things considered, the availability of high-resolution evidence regarding the Moon continues to be absent. The absence, once understood for what it suggests, begins to ring rather like thunder.

Even so, the absence continues in, for example, the form of the spy satellite, Clementine, launched by the U.S. Army in January 1994 from Vandenburg Air Force Base. According to many media reports, Clementine was equipped with various kinds of high-resolution cameras, including infrared and ultraviolet equipment.

Clementine's mission was successful, and the world awaited higher-resolution looks at the Moon via Clementine's sophisticated equipment.

The world awaited in vain—and thus we don't know what kind of bubble-gum wrappers are on the Moon. And so we will now turn our attention to the topic of what are referred to as "lunar anomalies."

By way of reference, one William H. Corliss set up an activity in 1974 called The Sourcebook Project with the goal of compiling "strange phenomena." In 1985, he published a substantial compilation entitled *The Moon and the Planets: A Catalog of Astronomical Anomalies,* which included 108 pages descriptive of Moon anomalies.

In this compendium, Corliss broke down the anomalies into eight major categories:

- ∅ The Moon's orbital anomalies.
- ∅ Lunar geology problems.
- ∅ Lunar luminous phenomena.
- ∅ The motion of lunar satellites.
- ∅ Anomalous telescopic and visual observations
- ∅ Lunar "weather."
- ∅ Lunar eclipse and occultation phenomena.
- ∅ The enigma of lunar magnetism.

Most of the categories listed above can be further subdivided. For example, lunar weather implies the existence on the Moon of an atmosphere. Lunar luminous phenomena include anomalies referred to as lunar transient phenomena (LTP) lights that come and go, and move around, and to green patches suggestive of vegetation, etc.

THE MATTER OF LUNAR LIGHTS

I n briefly turning to the matter of lights on the Moon, we encounter a situation that, simply put, is absolutely hilarious. To get into this, the compendium by William H. Corliss entitled *The Moon and the Planets* has been referred to in the preceding chapter. This compendium contains a section entitled "Lunar Luminous Phenomena," and begins with a brief Introduction.

In it Corliss points out that while the Moon has long been considered to be a "dead world," it nonetheless "exhibits a surprising variety of luminous phenomena."

He goes on to briefly discuss the Dead Moon Dictum long held by scientists and indicates "that luminous phenomena were seldom reported in the scientific literature" because "they couldn't exist" according to that scientific Dictum.

Corliss then points up that the "arrival of the Space Age brought the moon under detailed scrutiny; and both professional and amateur astronomers began reporting flashes of light, transient color phenomena..." and so forth.

To begin making the point of this chapter, it is necessary to elucidate the fact that the lunar light phenomena are NOT rare. They number in the thousands and some of them have been so robust as to have been observed by the naked eye.

Well, the arrival of the Space Age and the arrival in the vicinity of the Moon by Orbiters and later by manned Apollo missions ought to have brought further enlightenment about the nature of lunar light phenomena—especially those which have been reported as having motion and as moving about in definite ways.

Yet, the official silence since 1968 until today has been thunderous on this particular matter.

The year 1968 seems to have some kind of relevancy to all the matters discussed so far in this section.

What that relevancy consists of, however, is hard to pin down.

But that there is some kind of important tale in this regard can be clearly demonstrated.

The history of observations of lights on the Moon is rather long and, throughout it, accounts of the luminous phenomena were consistently logged in.

In large part, the accounts were ignored by various branches of the developing sciences.

But even so, up until the advent of the Space Age the luminosities could continue to be seen at rather frequent intervals.

Now, if one intended to colonize the Moon, which was publicized as the primary first goal of the Space Age, one certainly would like to know more about the lunar lights even if the mainstream sciences ignored them by clinging to the Dead Moon Dictum.

This factor really does need to be emphasized, because it IS a factor that has consistently been made invisible. If you were planning to send manned spacecraft to the Moon with the intention of building Moonbases, and if there were thousands of reports of sometimes awesome Moonlights, wouldn't you want to have some idea of what they were BEFORE sending your guys to the Moon?

At the very least, the larger telescopes might in some manner have been pressed into service. To think that something along those lines was NOT undertaken is just plain silly.

That the Space Age decision-makers were aware from the get-go of the lights was clearly established by NASA itself.

In 1968, a document was published entitled: Chronological Catalog of Reported Lunar Events (NASA Technical Report R-277).

The catalog documented 579 lunar events between 1540 and 1967, about 75 per cent of which referred to lunar "luminous phenomena." The remaining 25 per cent referred to phenomena consisting of Moon hazes, mists, fogs, and clouds that sometimes obstructed good telescope viewing of the lights, unless the lights moved beyond the foggy obscurations.

The authorship of the report was attributed to the joint efforts of four researchers, one each from the University of Arizona, the Goddard Space Flight Center, the Armagh Planetarium, and the Smithsonian Astrophysical Observatory.

Here it should be remembered that an "event" is actually a happening—i.e., something going on—in this case, going on with regard to the Dead Moon where "events" are not supposed to happen.

As indicated in its Introduction, "The purpose of this catalog is to provide a listing of historical and modern records that may be useful in investigations of possible activity on the moon."

A few lines later it states that "The catalog contains all information available to us through October 1967."

While this may be true as far as the document's four authors are concerned, the catalog most certainly DOES NOT contain all available information. Even after subtracting hundreds of reports known to be spurious, many more than 2,600 or so events rather than 579 events might have been included with complete justification.

For example, during the latter part of the nineteenth century, the Royal Astronomical Society in Britain recorded 1,600 lunar events over a mere two-year period by utilizing a 13-inch at the Royal Observatory at Greenwich.

So, the NASA catalog DID NOT contain all available information, but merely a selection of certain types of events from among it. In this sense, the NASA catalog contained data sifted from a great quantity of it.

The Introduction to the catalog also contains a section titled "Reports Omitted from the Catalog."

The first paragraph begins with the sentence "We attempted to eliminate all doubtful reports from this catalog." This presumes that the authors DID inspect a larger number of reports.

However, the first paragraph ends with the sentence "Careless reporting has been discovered in one case only." (This is identified as involving one "John Hammes and friends" of Iowa who reported seeing a lunar "volcano" on 12 November, 1878.)

If these two conflicting sentences contained in the same paragraph of the catalog seem at odds, it is because they ARE at odds.

One can only wonder why the NASA catalog was limited to only 579 events. And if this is considered, one might wish to wonder what kinds of historically documented reports WERE OMITTED from the NASA catalog.

Since the NASA catalog DID NOT include vast numbers of quite authentic lunar-event reports that could, even should have been included, the stated purpose of the catalog can be called into question. Certainly, the whole lunar truth, so to speak, is not reflected by the mere 579 events contained in the catalog.

And that leads us back to the subject of telescopes. The ordinary person with proper interest and some spare bucks can shop around for a very good 6-inch to 16-inch telescope which are commercially available.

Such a telescope can offer good information, provided Earth's atmosphere does not blur things too much, and that other night conditions are reasonably ideal.

If one now takes this particular factoid and compares it with the 597 lunar events listed in the 1968 NASA catalog, then it is revealed that the catalog contains not much more than those events that can be espied via an 8-inch, or at most a 16-inch telescope.

THIS particular factor is NOT mentioned in the introductory materials to the NASA catalog. The catalog thus seems to have the rather strange function of dealing only with lunar anomalies that can be seen from Earthside by smaller telescopes.

In any event, the publication in 1968 by NASA of the catalog of lunar events could easily lead one to expect that there would be a follow-on document AFTER the Moon was thoroughly examined up close by Orbiters and manned landings.

Certain craters on the Moon are rather notable for profusions of lights, and other anomalous phenomena as well.

Among these, and as duly noted in the NASA catalog, major among the craters exhibiting various kinds of lights are Plato, Aristarchus and Timocharis, to identify but a few.

Plato is famous for lights. It is about sixty miles across and has a floor that changes color. Its walls are quite high, but sometimes obscured by fogs and mists that bellow up and over them.

Many of the self-luminous "objects" are seen to move about. Others of them form geometric patterns, such as circles, squares, and triangles.

Sometimes lights have been seen emerging from smaller or near-by craters and moving to Plato, then descending the crater walls. In 1966, there were numerous reddish glowing spots shining out from Plato.

Beams and long-distance rays of light have also been observed. Space here does not permit further enumeration, and the interested reader might consult the bibliography for more detailed reading.

There is one other situation that should be noted. It regards where the Moon lights are most profuse and constantly reported versus where the manned Apollo missions landed.

If it were up to me, I'd probably set down one Apollo craft square into the middle of crater Plato. After all, it is sixty miles wide, and there appears to be a lot going on in it.

As it was to turn out, none of the American or Soviet spacecraft sent to the Moon landed anywhere near those special Moon locations.

Instead, all the landers were sat down in areas near the lunar Equator and into environments not noted for much except their lack of lunar activity. The observation above of course can only refer to craft publicly admitted as having been sent to the Moon.

LUNAR WATER- LUNAR ATMOSPHERE

D uring the spring of 1998, scientists announced their surprise at discovering water on the Moon. The news was spread far and wide not only in all major print media, but throughout the electronic bytes of the Internet.

However, as anyone familiar with the long history of Moon observations will realize, the 1998 discovery was not a discovery, but a very tardy admission of what had most certainly been known decades ago by official insiders.

One of the subtle importances of this is that where water is, an atmosphere can't be far behind. And where atmosphere is, sufficient gravity must be present to hold it down.

We are now somewhat far from the Dead Moon Dictum, which itself might be said to be dead.

But from the point of view of examining the dimensions of the lunar cover-up, it is worthwhile reprising certain aspects of the Dead Moon charade officially promulgated and maintained for so long.

As announced back in 1961, the major Space Age goal was to get to the Moon and colonize it for the purposes of erecting Moonbases as a step toward penetrating deeper into outer space. This obviously was a scientific and military goal.

Another major goal (now completely forgotten) consisted of economic opportunism. The Moon probably had useful resources that could be mined for the benefits of Earthside capitalistic ventures.

These two major goals were impressive ones that could, and did, arouse much enthusiasm for the lunar objectives, and which were extraordinarily costly.

Another factor needs to be considered. Once having gotten to the Moon, the conditions of habitat would be important.

It seems logical to assume that any lunar factor that might help ease lunar habitat problems would have officially been emphasized—in order, at least, to further increase the enthusiasm of taxpayers who were the ones actually funding the costly lunar expeditions.

For example, the lunar water "discovered" in 1998 was certainly present back in 1961 at the start of the lunar expeditions. The undeniable presence of the lunar atmosphere had been discovered, and confirmed, several decades before the advent of the lunar expeditions.

Indeed, beginning in the late nineteenth century, there came into existence a continuity of very competent observations about the Moon that clearly indicated any number of helpful lunar conditions that would ease the ultimate habitat concerns.

Simply put, water, atmosphere, voluminous caves (i.e., the negative mascons) to reside in, minerals, etc. What could be more wonderful with regard to resolving at least some of the lunar habitat problems?

Beyond all this, then, and considering the excellence of human genius and ingenuity, the only remaining REAL problem, if there WAS one, would consist of what might be called "occupational hazards" of a kind Earthside genius and ingenuity might have REAL problems dealing with.

There now remain two additional factors to be carried in mind.

First of all, with the advent in 1961 of the active measures segment of the lunar expeditions, the official stance regarding the Moon elected to reinforce the negative Dead Moon Dictum.

Doing so broke the continuity of earlier observations about the Moon, observations that positively foreshadowed relatively easy colonization overtures.

Second, the lunar fly-bys, Orbiters, Landers and manned Apollo landings lasted approximately eleven years—after which the two great Earthside superpowers, the United States and the Soviet Union, completely abandoned treks to the lunar satellite.

After this abandonment, official propagandizing sought to stimulate scientific and public enthusiasm for the proposed wonders of space stations that were to orbit Earth, not the Moon.

All things considered, it is possible that the combined genius and ingenuity of the two superpowers encountered certain lunar occupational hazards that were difficult to deal with—deterrents which must have been of such impressive quality that they disrupted the best laid plans for the projected lunar conquests.

So, the two superpowers, technically still in Cold War with each other, decided to cooperate, of all things, on building an Earth-orbiting space station.

Whatever the combined and undoubtedly enormous costs of this particular effort, orbiting space stations will always be quite fragile as compared with Moonbases—and which might have been routinely achieved by the early 1980s.

In order to help construct some kind of format regarding the lunar cover-up, we can benefit from backtracking to the beginning of the twentieth century to W. H. Pickering, a Harvard professor and an accepted authority on the Moon.

As he noted in his highly professional book *The Moon* (1903), the view that the Moon is a dead unchanging world is so widespread and firmly rooted in minds that not only the general public, but the astronomical world as well, are united in this unanimous opinion.

He goes on to state that the unanimous opinion is based on the most inadequate negative evidence.

He refers to hundreds (or thousands) of telescopically observed Moon changes and indicates that the only plausible explanations of the changes is that they involve the presence of air and water.

He then went on to identify a situation that most are not aware of, and which involves selenographers—those who scientifically study the physical features of the Moon.

"The arguments on the two sides of the case are extremely simple. The astronomers who are not selenographers declare that there is no atmosphere or water on the Moon, and that, therefore, changes are impossible. The selenographers' reply is simply that they have seen the changes take place."

However, in spite of many great selenographers, the Moon thereafter officially remained without atmosphere for about ten decades—even though two noted and competent astronomers published evidence to the contrary. Both astronomers were able to conclusively document the existence of the lunar atmosphere based on 150 years of scientific records and telescopic studies.

One of those astronomers was M. K. Jessup, who had taught astronomy and mathematics at the University of Michigan and went on to build the largest refracting telescope in the southern hemisphere.

His book about the Moon was published in 1957 (please note the date) under the title, believe it or not, *The Expanding Case for the UFO*—and which presumably got him in deep dodo with several layers of Earthside cryptocracy.

The other courageous writer was V. A. Firsoff who was acknowledged at the time to be the top scientist and authority on the Moon. His 1959 book (please note the date) was entitled *Strange World of the Moon.*

Both of these books presented evidence (not easy to contradict scientifically) of a lunar atmosphere and the high probability of regional water and vegetation.

The only real result of these two books was that they were quickly caused to be out of print, etc., and became hard to locate as they still are today.

The existence of the atmosphere, however, could be determined by the way stars are "occulted" when the Moon passes over them.

If the Moon had no atmosphere, they would simply vanish instantaneously as the lunar body moved in front of them. However, if there is a gaseous layer, an atmosphere, then the stars begin to flicker before they pass behind the Moon.

Stars flicker as they become occulted by the Moon—leading to the general conclusion that the lunar atmosphere is about three miles thick, and more dense near the lunar surface. The lunar atmosphere also provides sufficient friction for "flashes" as small meteors become incandescent upon passing through it.

When the first manned mission (Apollo 11) landed, the astronauts planted a flag and filmed this triumphant event.

Shortly after the astronauts planted the flag, and while the on-site TV camera was running, an errant gust of wind came along and billowed the flag outward.

William Brian, author of *Moongate: Suppressed Findings of the US Space Program* (1982), obtained a copy of the film. It showed that the astronauts were not close to the flag when it started waving. Being nearer the camera, both ran to block its lens with their arms and hands.

NASA could not be encouraged to comment. But when flags were thereafter planted on the Moon, they were rigged with wires and mesh so that they stayed rigid at all times. This incident was ultimately forgotten. But wind needs the presence of atmosphere in order to blow flags this way and that.

The presence of an atmosphere on the Moon "permitted" the real existence of the lunar mists, fogs and clouds which many hundreds of viewers began to notice after about 1733 (when proper telescopes began to be invented.)

The nineteenth century was particularly rich regarding lunar atmospheric phenomena when the art of building large refracting telescopes reached a zenith benchmark.

In any event, all lunar phenomena that couldn't be fitted into the dead Moon dictum were excluded from mainstream scientific workings. With the dawn of the Space Age, however, something along the lines of a governmental space agency was required.

In the United States, this need evolved into the National Aeronautics and Space Administration (NASA) which came into official existence on October 1, 1958.

In official literature, NASA is described as a civilian agency of the U. S. Federal government with the mission of conducting research and developing operational programs in space exploration, satellites and rocketry.

One of the first major objectives of NASA's space exploration mandate was in two parts: (1) to get to the Dead Moon; and (2) to take possession of it and colonize it with Moonbases before the fearsome Soviets did.

Now, this Dead Airless Moon target was the same Moon that was KNOWN by 1958 to have mists, fogs and clouds. Mists, fogs and clouds clearly refer to some kind of lunar weather, even if it is not akin to Earthside weather.

It thus needs to be emphasized that before 1958, thousands of historical reports existed regarding lunar anomalies (including weather phenomena).

The sum of these thousands of reports clearly established that the Moon was NOT dead and deprived of activity.

It is, of course, quite ridiculous to think that NASA personnel did not examine in excruciating detail this mass of historical reports.

As it was, the existence of a "weak" lunar atmosphere was finally "discovered" in 1997. But there was little commentary about what the atmosphere permitted—such as the vegetation here and there on the lunar surface, to say nothing of the clouds, etc.

MOON-STYLE OCCUPATIONAL HAZARDS

I t is to be expected that layers of secrecy might be employed regarding lunar matters in order to gain military or economic advantages. After all, it is rather a permanent aspect of human nature to secretly strive for advantages in both arenas.

An analysis of the official secrecy regarding the Moon reveals that over decades covert methods were employed to hide the existence of lunar water, atmosphere and several other ameliorative phenomena made possible by them.

One of these additional phenomena is that seeds sprout and grow quite well in lunar soil which is rich in nutrients. This indicates the possibility that lunar colonizers from Earth-side might somehow grow their own food supplies on the Moon.

All of these are benign phenomena, suggestive of easier habitat factors than if the lunar satellite WAS actually dead, dry, airless, and formidable.

In this regard, several unofficial sources have been undertaken to assess these lunar conditions outside of the officially maintained cover-up.

Fred Steckling and Daniel K. Ross (see bibliography) have provided logical and well-reasoned analyses based on official lunar photos. Their estimates more or less conclude that dwelling on the Moon would resemble dwelling in the higher Andes or Himalayan mountains, but that healthy and vibrant Earthsiders would acclimate quite well to such conditions.

In a push, natives born in the high Andes or in Tibet could be recruited for Moon habitation and trained for military or economic purposes.

What an analysis of the official secrecy thus reveals is that the secrecy was never really necessary in the first place—if only the elements discussed above were involved.

Indeed, those benign elements would have benefited the Moon missions if they had been publicly disclosed. For one thing, venture capitalists certainly would have invested in stratospheric funding.

Coming directly to the point, the cover-up secrecy does not itself reveal why the two Earthside superpowers abandoned the Moon after having instituted such a vigorous, costly and highly publicized start to get there.

On October 12, 1954, about six years before President Kennedy announced the great American effort to place a man on the Moon, an astronomer was utilizing the telescope at the Edinburgh Observatory to examine the Moon.

He was able to observe "a dark sphere travel in a straight line from the crater Tycho to the crater Aristarchus "both of these craters otherwise demonstrating numerous light phenomena of one kind or another.

The distance covered by the sphere took "a period of twenty minutes, and this roughly calculated to a speed of nearly 6,000 miles per hour."

In September of the same year, a similar object had been sighted by two men using a 6-inch telescope. For over forty minutes they watched it leave the northern area of Mare Humboldtianum and move upward out of the lunar atmosphere into space.

It is important here to recall that small-scale telescopic resolution requires lunar objects to be very large scale in order to be perceived. The spherical objects must have been quite large, say about one to four or five miles in dimension.

Indeed, many reports of "anomalous black bodies" crossing the surface of the Moon are on historical record, some of them casting their shadows on the lunar surface as they expeditiously move along in what can only imply directed flight. Additionally, similar but much smaller self-glowing objects have been espied flying in formation in and out of this or that crater.

None of these objects was included among the 579 anomalous phenomena listed in NASA's 1968 *Chronological Catalog of Reported Lunar Events.*

Even so, such craft may have had something to do with why the

Americans and Soviets decided to cease putting their men on the Moon and transfer their efforts to space stations closer to Earth.

For all their dramatic excellence, manned lunar Landers must be considered as quite wimpy compared to a rather stalwart craft of about four miles wide and capable of achieving 6,000 miles per hour.

Indeed, such craft, or at least their personnel, might constitute something of a lunar occupational hazard—at least in terms of occupying the Moon.

Two NASA photos showing extraordinary crafts seem to have escaped censorship and airbrushing.

The first of these comes from Apollo 11, when, in July, 1969, its camera inadvertently captured a really neat and clear photo of a glowing, cigar-shaped object close to the lunar surface. Since the photo reveals a vapor trail, the craft must have been traveling somewhat within the lunar atmosphere. (NASA photo No. 11-37-5438.)

In July, 1972, the Hasselblad camera of Apollo 16 recorded yet another cigar-shaped object. This object was quite large. It seems to have been somewhat glowing white (ionizing the atmosphere directly next to it) but was close enough to the lunar surface to cast its equally elongated shadow. (NASA photo No. 16-19238.)

Even after the two Earthside superpowers did not return to the Moon, telescope enthusiasts around the world have continued to observe and document airborne vehicles traveling in proximity to the lunar surface, above its atmosphere, or departing or arriving from space.

Especially impressive in this regard are the Japanese enthusiasts who have captured telescopic TV footage of the "anomalous" phenomena, which has been viewed world-wide.

There are multitudes of reports of this kind. And interested readers should consult the bibliography for sources that review the structures more in-depth.

The bigger question is, WHAT the larger, official objective of the Earthside cover-up seems to be preoccupied with covering up.

It is apparently of such a nature as to make secretive Earthside powers think twice about colonizing the Moon, or perhaps even returning to it until things get better sorted out.

Whatever it is, it is not lunar water, atmosphere, minerals, or lights per se. Why would anyone want to cover up water on the Moon? But there might be "hazards" utilizing the water there, and the "hazards" might need to be covered up—especially if Earthside powers didn't know how to deal with them.

One possible speculation about the real nature of the cover-up is that someone on the Moon kicked Earthside ass and seems to have "suggested" that it not come back.

Any reader truly interested in the actuality of the cover-up might wish to track down a book which by now is truly rare.

It was first published in 1978 in France, then translated into English. It was written by one Maurice Chatelain, who in 1955 came to the United States from (then) French Morocco.

His book was entitled *Our Ancestors Came from Outer* Space, but it includes quite a number of factoids such as: "When Apollo 11 made the first landing on the Sea of Tranquility, and, only moments before Armstrong stepped down the ladder to set foot on the moon, two UFOs hovered overhead."

Chatelain later commented that: "The astronauts were not limited to equipment troubles. They saw things during their missions that could not be discussed with anybody outside NASA. It is very difficult to obtain any specific information from NASA, which still exercises a very strict control over any disclosure of these events."

While many have laughed, or remained mute, regarding Chatelain, he constitutes a particular problem against which the cover-up cryptocracy can do little else than say: "No comment."

For Maurice Chatelain had been placed in charge of designing and building the Apollo communication and data-processing system for NASA.

Earlier, he had been in charge of engineering new radar and communications systems for Ryan Electronics in the late 1950s. He had received eleven patents, including an automatic radar landing system used in the Ranger and Surveyor Moon flights.

In other words, Maurice Chatelain was a cryptocracy INSIDER, and to some degree at least he must have known what he was talking about.

Two more secrecy tidbits from Chatelain: It seems that all Apollo and Gemini flights were followed...by space vehicles of extraterrestrial origin.

Rumors within NASA were that Apollo 13 carried a small nuclear charge. It was designed to be set off on the Moon for seismic testing. But the craft barely managed to return to Earth-after being disabled by a UFO that seemed intent on protecting some Moonbase "established there by extraterrestrials.

There is much more that could be said regarding the Moon. But I'll end this section with the observation that back in 1975 when Mr. Axelrod first contacted me, he knew all of this—and probably MUCH more.

EARTHSIDE TELEPATHY

VERSUS

SPACESIDE TELEPATHY

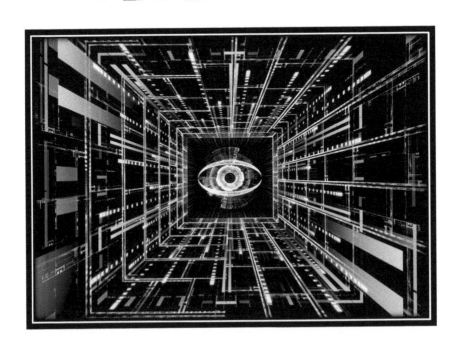

INFORMATION PACKAGES KEPT APART

It can easily be established that some kind of official cover-up regarding the Moon was put in place long ago and has been maintained all along. But the nature and dimensions of the cover-up are a lot less clear, and it is this lack of clarity that introduces confusion into the situation.

Contributing to the confusion is that when some kind of penetration into the cover-up does take place, the official organizations involved simply march on anyway, and the cover-up remains largely undisturbed.

There have been very many penetrations of this kind, with the information downloading from them ending up in books. Some of the books, of course, are somewhat over-baked or slightly hysterical. But many of them are calmly organized and quite well documented.

The revelations by Maurice Chatelain earlier referred to are but one example. Chatelain was a strategic NASA insider, and must be accepted as having been well-informed.

Many encounters with Spacesiders must therefore be accepted as factual, and those who can tolerate the implications have done so.

It is helpful along these lines to briefly turn attention away from what is being covered up, and to give some attention to how cover-ups are made to work.

Of course, the best possible kind of cover-up has to do with installing very tight secrecy around something. This can be successful, to the degree that outsiders have no awareness that anything is going on.

But the Moon cannot be surrounded within such tight secrecy—because there it is, in the sky for everyone to see.

So, the Moon itself can't be covered up. But activities on it can. This is to say that if there are lunar activities, and if the cover-up involves those activities, then the best way to institute the cover-up is to establish the Dead Moon Dictum, to give the Dictum cultural and scientific authenticity, and to teach the Dictum in schools.

Then when people look up at the Moon, there it is—and perceptions of it are surrounded in the disarming glory of the Dead Moon Dictum. Aside from an occasional light on the lunar surface, people can't see anything else.

However, some people purchase telescopes. These are not very high-resolution telescopes, to be sure. But through them, many of them report seeing things that can't exactly be fitted into the Dead Moon Dictum. And so the problems of managing the cover-up become a little more intricate.

In the first instance of this, however, there is an easy solution at hand. This simply consists of officially trashing those telescopic observations of the Moon that can't be fitted into the Dictum.

Thus, outsiders who purchase telescopes can see whatever they do, but this really doesn't matter because the big fists of mainstream officialdom are what count.

In the second instance of managing the cover-up, various lunar activities espied outside of official auspices can either be explained away or simply, and expediently, ignored. For example, most of the books listed in the bibliography are simply ignored—at official, mainstream levels, anyway.

Returning to the bigger picture of the lunar cover-up, if one patiently sifts through all available information, it might appear that the cover-up has only to do with natural phenomena of the Moon—with the existence of lunar atmosphere, water, vegetation, etc., however minimal those might be.

Indeed, in the face of evidence to the contrary, the existence of lunar atmosphere and water (only recently admitted as existing) were denied for a long time, while the denial itself was clearly part of the complicated cover-up.

But one might wonder WHY natural lunar phenomena needed to be covered up, or denied. Under usual circumstances, the world would have been enchanted to find that the Moon was not Dead.

Further, this wonderment needs to be compared to the intensity of the cover-up-for official denials of anything strange or surprising about the Moon have been enormous and enduring, almost in an over-kill way.

As a general rule of thumb, natural phenomena are not usually covered up (at least for very long) unless there is a very good reason for doing so. And so, many have wondered if something beyond natural lunar phenomena is the actual objective of the cover-up.

In the light of the above, cover-ups do benefit from confusions proliferated and maintained at official levels, in that the on-going efficiency of the cover-up can be maintained by proliferating the confusions.

Indeed, disinformation experts have evolved elaborate and efficient methods for concealing something by surrounding it with confusions.

One way to help maintain confusions is to keep separated various kinds of information packages that are entirely relevant and even necessary to each other.

Such information packages need to be combined, or juxtaposed, in order to make sense out of what otherwise remains a confusion that can be capitalized upon for the delightful benefit of cover-up.

One method of digging into the nature and dimensions of a cover-up is somehow to simplify the confusions in order to see what can be seen beyond them.

This helps different information packages to become somewhat visible, or to stand out more clearly. It also helps in the process of discovering what information packages are remaining invisible or untouched, and which are not being factored into consideration.

Sometimes erecting a simple chronology of events and developments is useful along these lines. Doing so helps things fall into consecutive place.

Although some books downloading good, even provable information relevant to the cover-up contain a wealth of meaningful factoids, the whole of the downloading is none the less heaped and twisted together in such a way as to convolute the facts rather than aligning them.

The *New Columbia Encyclopedia* (Fourth Edition, 1975) has a rather nice entry for "Space Exploration," and from it I have derived the brief chronology that follows.

I have laced the chronology with certain Comments that belong with it timewise, but which of course are not mentioned in the Encyclopedia.

The entry begins with a definition of space exploration, which consists of the investigation of physical conditions in space and of stars, planets, and their moons through the use of artificial satellites, space probes, and manned spacecraft.

It is further indicated that although studies from Earth using optical and radio telescopes had accumulated much data on the nature of celestial bodies, it was not until after World War II that the development of powerful rockets made direct exploration a technical possibility.

It now needs to be pointed up that the Encyclopedia refers to optical telescopes via which it can be presumed various kinds of information packages (about the Moon, for example) were acquired. However, as has been outlined in Chapter 16, the matter of the telescopes has never surfaced as a straightforward one—in that the larger telescopes have been sequestered under official control, and this on a world-wide basis.

COMMENT: Common sense tells us that large telescopes were utilized to spy on the Moon as early as the 1920s. Yet, it should be noted that no downloading of information from these telescopes has ever descended into public cognizance. It is therefore possible to assume that elements of the cover-up had emerged as early as the 1920s.

What was discovered and determined via the larger optical telescopes has become a permanently missing information package.

October 4, 1957. The USSR launched the first Earth-orbiting artificial satellite, Sputnik I. The dormant U.S. program is thereby spurred into action, leading to international competition, popularly known as the "space race."

COMMENT: The "Space Race" for what? As early as 1961, it had been clearly established in the open media as consisting of who was to be the first to acquire supremacy in space, with particular emphasis on colonizing the Moon.

In fact, the American effort was galvanized into feverish activity because of the fear that the Soviets would acquire the supremacy.

This particular factoid is omitted from the Encyclopedia— and is likewise omitted from all materials published AFTER the United States desisted from further Moon explorations in December, 1972.

January 31,1958. Explorer I, the first Earth-orbiting American satellite is launched.

Having indicated this much, the Encyclopedia goes on: Although Earth-orbiting satellites have by far accounted for the great majority of launches in the space program, even more information on the Moon and other planets, and the Sun, has been acquired by unmanned space probes and manned spacecraft.

In the decade following Sputnik I, the United States and the Soviet Union between them launched about 50 unmanned space probes to explore the MOON.

The first probes were intended either to pass by close to the Moon (flyby), or to crash into it (hard landing.)

- ∅ September 1959—The USSR Luna 2 made a hard lunar landing.

- ∅ November 1959—Luna 3 took pictures, for the first time ever, of the Moon's far side.

- ∅ February 1966—Luna 9 achieved the first lunar soft landing.

- ∅ April 1966—Luna 10 orbited the Moon.

- ∅ Both Luna 9 and 10 sent back many television pictures to Earth.

The Encyclopedia goes on to state that American successes generally lagged behind Soviet accomplishments by several months, but provided more detailed scientific information.

COMMENT: A rather strange factor now needs to be interjected regarding the "more detailed scientific information." It is permissible to assume that at least some detailed information might vastly change the stereotyped Dead Moon picture. And indeed, many scientists not only acknowledged the existence of such information but wrote and published papers regarding it. Yet nothing that seriously conflicted with the Dead Moon Dictum was officially acknowledged to the public or integrated into standard academic or media sources.

The Encyclopedia continues: In the U.S. program, the early Pioneer launches were largely failures, as were the first five launches of the Ranger series, which attempted semi hard landings of rugged instruments. Subsequent Rangers carried only television cameras and impacted at full speed.

Beginning in July, 1964, Rangers 7, 8 and 9 transmitted thousands of pictures, many taken at altitudes less than 1 mile just before impact and showing craters of only a few feet in diameter.

July, 1966. Surveyor 1 touched down. In addition to television cameras, it carried instruments to measure soil strength and composition.

COMMENT: Public cognizance was duly informed about lunar soil strength and composition. Whatever else the television cameras might have filmed was never commented upon.

August, 1966. The United States successfully launched the first Lunar Orbiter, which took pictures of both sides of the Moon as well as the first pictures of Earth from the Moon's vicinity. The primary mission of the Orbiter program was to locate suitable landing sites for Apollo, the manned spacecraft program.

COMMENT: The landing sites selected for the Apollo missions turned out to be some of the most featureless, arid, desolate locations on the Moon, and were in proximity to the lunar Equator. There is no publicly available evidence that either the Soviets or the Americans even sent a television camera into, say, the craters Plato or Aristarchus—otherwise known for copious lunar anomalies, some of which are described as being quite lush.

Between May *1966* and November *1968,* the United States launched seven Surveyors and five Lunar Orbiters to photograph and map the Moon.

COMMENT: However, it seems that none of the photographing and mapping included any areas noted for their anomalies.

Also in 1968, NASA released its *Chronological Catalog of Reported Lunar Events.* The strangeness of this catalog has already been discussed in Part Two. "Lunar events," of course, should be read as "lunar anomalies." The catalog listed multitudes of lights and other phenomena taking place in certain quite large lunar craters.

NASA never published any follow-ups on the lunar anomalies, even though it might have done so by virtue of the enormous amounts of information derived from the Surveyors and Orbiters.

When the manned Apollo crafts DID finally arrive at the Moon, all of the locations selected for the touchdowns were far distant from any of the sectors that had always yielded high incidence of anomalous activity.

The following manned Apollo craft landed on the Moon:

Ø   July 20, 1969: Apollo 11.

Ø   November 19,1969: Apollo 12.

Ø   February 5,1971: Apollo 14.

Ø   July 30,1971: Apollo 15.

Ø   July 30,1971: Apollo 16.

Ø   December 11,1972: Apollo 17.

The Soviets sent to the Moon the following unmanned Luna crafts:

Ø   September 20, 1970: Luna 16.

Ø   November 17, 1970: Luna 17.

Ø   February 21,1972: Luna 20.

Ø   January 16,1973: Luna 21.

Ø   August 16, 1976: Luna 24.

Regarding the line-up above of the American and the Soviet expeditions to the Moon, the Columbia Encyclopedia states:

"Until late 1969, it appeared that the USSR was also working toward a manned lunar landing....After Apollo 11, however, the USSR apparently abandoned the goal of its own manned lunar exploration....After the Apollo program, the United States continued manned space exploration with Skylab, an earth-orbiting space station that served as workshop and living quarters for three astronauts."

The trusty Encyclopedia does not come directly out and SAY that the United States abandoned its own Moon excursions after Apollo 17 in December 1972.

But such is the direct implication.

Thereafter, public attention was directed to the awesome potentials of Skylab, and to space craft launched, in 1971, to the planet Mars. The possibility of Moonbase quickly, and too quietly, receded from public cognizance.

Indeed, few were aware that manned excursions to the Moon had ceased. I, myself, until sometime after my encounter with Mr. Axelrod in 1975, did not notice that such excursions had ceased.

If the brief chronology outlined above is taken at face value, it appears to hold water. But if one attempts to identify what information packages are missing from it, then the chronology becomes quite wobbly.

One of these missing information packages might consist of a companion chronology of UFO activity. As it turns out, the Earthside space effort is NEVER discussed within the contexts, or the chronology of, UFO activity—and which, to all apparent purposes, is Spaceside space activity.

For clarity, the information package of Earthside excursions into space is never discussed alongside information packages of Spaceside excursions into Earthside space (i.e., our planet and its Moon).

I'll now reduce the above Earthside space excursions into a quite simple chronology.

∅ As of about 1958, Earthsiders proposed to get to and colonize the Moon with Moonbases.

∅ Earthsiders first got to the Moon with TV cameras and

sensitive instruments, and between 1969 and 1972 physically landed on it.

Ø After achieving this much, Earthsiders shifted focus away from the Moon.

Ø The Moon was never officially heard of again until the early 1990s when the U.S. Army launched project Clementine—a lunar orbiter with three kinds of cameras capable of high-resolution of the kind that can read from space a bubble-gum wrapper in a New York City gutter.

High-resolution photos from Clementine's awesome spy-in-the-sky cameras have not been released—although many photos of lesser resolution have been published.

Why the Moon was abandoned is a reasonable question, especially after the earlier enormous enthusiasm and billions of dollars poured into attempts to get there.

Well, Earthsiders got there. But they never went back.

In the sense of all of the above, then, there are several rather large information packages flopping around.

Why we never went back to the Moon is certainly one of them. In this sense, why one does NOT go someplace if one builds the costly equipment to get there is, after all, something to be wondered about.

THE PROBLEM OF INTELLECTUAL PHASE LOCKING

It is quite easy to assume that there is an Earthside cover-up regarding some kind of Spaceside factor that seems to necessitate the cover-up, at least in the minds of those insider officials who might have access to ALL relevant information packages.

On the other hand, just outside the margins of the cover-up an entire counter-cover-up industry has come into existence embodied in thousands of books and articles about WHAT the cover-up is covering up.

In this way, a very complicated relationship has developed between the cover-up forces and the counter-cover-up revolutionaries. It is, of course, easy to identify the obvious factors of this complicated relationship.

However, if the issue involved only the obvious factors, then the cover-up could not work for very long.

As it has transpired, the cover-up might have had quite fine legs to stand upon during the 1950s. As time went on, though, the cover-up has literally become quite flimsy.

But it continues in power anyway. WHY it continues in power is difficult to articulate.

One can really begin to wonder if, in its bigger-picture sense, the cover-up is covering up the actuality of Spaceside activity of numerous kinds—one kind being the vivid and very frequent appearance of UFOs that are visible world-wide to Earthsiders.

The Earthsiders who chance to witness UFO activity are, of course, outside of direct cover-up control parameters whose insider personnel organize and promulgate intellectual reasons by which the UFO witnesses cannot have seen or experienced what they did.

The whole of this certainly seems akin to a rather silly but remarkable revolving door which both spits out and chews up information, among other things.

It also seeks to distort and to destabilize the reality confidence factor not only of the thousands of witnesses, but of the general Earthside populations as well.

Thus, the counter-cover-up industry seeks to reveal the facts about WHAT the cover-up is covering up—and in this regard copious amounts of delicious data have been presented for those interested in tracking it down.

In this sense then, the counter-cover-up enthusiasts work to demobilize the cover-up by attempting to put the lie to the cover-up. However, this in turn means putting the lie to the official echelons which have promulgated the cover-up—and continue to do so regardless of the availability of counter-cover-up information.

In this way, a powerful dichotomy has come into existence. DICHOTOMY is defined as a division or the process of dividing into two—especially into two mutually exclusive or contradictory groups, or into two contradictory information sets.

Although the great uninitiated masses might not understand it very well, there is a famous dictum along these lines called Divide and Rule.

At first sight, one might think this has nothing at all to do with the cover-up and counter-cover-up fiasco, but there are several elements of Divide and Rule that can become apparent if one patiently constructs a larger picture of what is involved.

Rulership through dividing requires that the dividing first result in rather hefty and perpetuating confusions—behind which, and through which, the rulership can be effective.

As it is, the counter-cover-up enthusiasts tend to focus on the extraterrestrial details which are being covered up, many details of which can be completely documented as fact.

The cover-up forces continue to "rule" anyway—largely, it might seem, because the facts apparently don't matter on the one hand, while on the other hand the resulting confusions seem to aid and abet the cover-up.

This kind of rather astonishing situation suggests that if the cover-up was to be submitted to legal procedures and dragged into court where not insignificant amounts of evidence would be considered, the most probable outcome might be that the cover-up would undergo indictment.

The cover-up, however, is not submitted to legalistic inspection. Instead, it is "submitted" to science, and to scientific oversight—and the whole of which not only pompously comes down with a mainstream bang on the side of the cover-up but can be seen as giving sustenance and artificial life-support to it.

As but one easily accessible example of this, the atmosphere of the Moon had been identified as early as the 1920s, and, as well, identified by prominent scientists of that decade.

With the advent, however, of the Earthside space race to colonize the Moon, the prevailing wisdom of Science obstinately downloaded into public cognizance the authenticity of the airless Dead Moon Dictum.

The airless Dead Moon Dictum was maintained in place by SCIENCE until 1997, when Science thence "discovered" the lunar atmosphere.

With this announced 1997 discovery, the direct meaning is that the Apollo astronauts DID NOT land on an airless Moon as vividly proclaimed—for if the lunar atmosphere was there in 1997, it was certainly also there back in 1970, and was certainly there when it was first espied and identified back in 1920, or earlier.

Previous to the 1997 "discovery," however, the counter-cover-up workers had published in very numerous books enormous quantities of Moon atmosphere evidence, the authenticity of which should automatically reinstate the many authors as having been factual all along.

As it was, an information package regarding the existence of the lunar atmosphere had been available all along. And its existence had been documented well before anyone actually thought of colonizing the Moon before the Soviets did.

The existence of the lunar atmosphere information package was scientifically denied, and the scam of the airless Dead Moon Dictum package was, as it now must be said, FOISTED into public cognizance and acceptance.

We could circulate through all of the obvious and subtle issues involved by now. For example, the reason WHY the lunar atmosphere was announced in 1997, when it could have been announced in 1958 when the lunar conquest program got underway.

Since the 1997 announcement, clearly downloads not from science per se, but from the cover-up strata that incorporate science, there must be a good reason for it. Whatever it is, however, remains opaque.

In any event, an effective cover-up scam requires a wide latitude of erected confusions in order to succeed.

The confusions have to do not with facts, but with how Earthsiders think of them, or how they can be encouraged to think of them. Earthsiders, of course, think by processing packaged information rather than by processing random data that has not yet been encapsulated into a packaged form.

The distinction between random and packaged information is that the former has not achieved much in the way of meaning, while meaning has been attributed and assigned to the latter. As it is, Earthsiders are not all that interested in data that, to them, do not mean very much.

Thus, more precisely defined, packaged information is meaning-managed information—because of which, and out of which, intellectual phase-locking among biological separate individuals can take place.

More simply put, if groups of Earthsider individuals can be brought, one way or another, into agreement about the meaning of something, then their communal intellectual processes will phase-lock with each other.

Groupthink can then be formed with respect to this or that information package resulting in that intellectual phenomenon earlier referred to as mindsets.

The most immediate result is a kind of group-mind thing.

If one takes time to consider the logical emanations of this, the basic purpose of the Space Age cover-ups has NOT simply been to deny certain factors in the face of evidence that supports them.

It is far more likely that a concerted, and rather successful, attempt was undertaken with regard to TWO principal functions:

1. To increase rather than decrease space age confusions, so as better to promulgate and rule via disinformation packages.
2. To erect and reinforce a particular kind of planet-wide intellectual phase-locking that is data deficient with regard to the meaning not of Earthside affairs, but with regard to the meaning of Spaceside activities.

Anyone who has read any of the counter-cover-up materials will recognize that the cover-up has been exceedingly successful with regard to (1) above.

But although some few might intuitively respond to (2) above, the enormous hubbub of (1) has completely obliterated the *MEANING* of Spaceside activities—and which is so low as to be nil.

The most probable way (2) has been achieved is by keeping separate various kinds of information packages—which, if integrated, might contribute to at least some discovery regarding such meaning.

As it is, Earthsiders are completely malleable regarding (2) above, in that by historical habit they intellectually phase-lock on limited numbers of information packages—and eject those that don't fit into the phase-locking.

Thus, the whole (so to speak) of all possible information packages is kept broken apart. And THIS, of course, is quite convenient to the time-tested Divide and Rule procedure.

It should be pointed up here that this understanding is not novel or original to this writer. Others have similarly identified it, albeit under the different nomenclature of the individual's "local realities" versus "non-local realities."

The latter, of course, refers to realities that are larger and more encompassing—even to the degree of being universal in their contexts.

This is to say that Earthsiders do not think outside of Earthside local realities.

This is further to suggest that the realities of Spacesiders might not fit into ANY recombination of Earthside information packages—and especially so IF Earthside intellectual phase locking is deficient with regard to any Spaceside realities except those officially admitted to by Science.

And here it must be pointed up that the Earthside Sciences are focused only on the physical aspects of whatever is in space.

While almost all Spaceside realities constitute nothing less than a complete mystery, there is at least one of those realities that can easily intellectually phase-lock with one Earth-side reality.

But in attempting to elaborate on this, we will encounter what amounts to an Earthside reality that in itself is deeply mysterious.

THE "TELEPATHIC" CONNECTION?

My first encounter with Mr. Axelrod in the underground place took place in 1975. This was approximately three years after Apollo 17 had visited the Moon, after which U. S. interest in colonizing the lunar satellite seems to have evaporated.

I, however, was not then aware of the evaporation—believing, as most did, that the lunar conquest was somehow on-going.

Likewise, I thought the airless dead Moon was in fact just that. I did not begin to accumulate the information in Part Two until the mid-1980s.

It will be recalled that Mr. Axelrod and I discussed telepathy, and that he ultimately asked me to jot down my thoughts in that regard. At the time, I didn't give this request much thought, more or less thinking that it was just a natural part of discussing ESP in general.

I don't remember exactly what I wrote down, but I do remember that Mr. Axelrod's face lost its perpetual conviviality at this topic, his lips drawing into something of a thin line.

I assumed that the Axelrod affair was completely over and done with until the inadvertent event in Los Angeles which included the overly sensual female, my ultimate goosebumpish response to her, and the sighting of the twins.

Without the sighting of the twins, I would certainly have attributed any ET factor to my imagination—largely because what Earthsiders cannot explain we allocate to that widely-shared intellectual phase-locking called "fantasy."

The next thing that happened was Mr. Axelrod's telephone call to me in Grand Central station (of all places)—during which Axelrod pumped me for information about whether the female had psyched me out.

When thereafter my somewhat overworked synapses had cooled back into some kind of working order, it slowly began to dawn on me that Axelrod was actually in an uproar over what we Earthsiders refer to as telepathy—not on my part, but on the part of the female.

The fact that Axelrod was quite serious with his inquiries seemed to imply that the female, or extraterrestrial, had a SERIOUS type of telepathy capable of something dire. A type of telepathy plus, as one might suppose, a type which beyond being a channel for information exchange might also achieve something along the lines of mind bending and fried brains.

However, it was only Mr. Axelrod's emphatic concern that shifted this into the fact that it was later to become in my own appreciation of things.

He knew *They* existed, that at least some of them were dwelling Earthside, and that They WERE in possession of telepathy plus.

Thereafter, when time permitted, I expended effort to not only understand more about telepathy, but to try to think of it outside of the standard intellectual phase-locking typical of the modern age.

Although I don't remember much of what I noted down for Mr. Axelrod about telepathy in 1975, I would have included certain conclusions I had already concretized.

I would have indicated that telepathy must be inherent in our species, and not simply a matter of certain unique individuals seemingly specially gifted with it. After all, some of the most well-documented cases of telepathy involve spontaneous experiences at the mass level.

What I would not have included was an understanding I came to much later. This involves the matter of the nature of consciousness as will be discussed—not individual consciousness, but Consciousness as a universal premise and life force.

As to telepathy plus, I didn't have much reality on what THIS would consist of—until in 1989 when I began in-depth studies of CHI GONG. THEN I began to have some idea of what telepathy plus WOULD consist of.

The fact that telepathy plus IS possible for Earthsiders is the fundamental fact that has caused me to write this book.

Although my knowledge about telepathy has a good way yet to go, I now know much more about it—about both what it must consist of, and what it doesn't consist of.

In accumulating this information package, it became possible for me to make the following and quite basic observation, an observation that is easy enough to substantiate.

Telepathy is the most forbidden element of Earthside consciousness. Indeed, so forbidden that Science would rather accept reincarnation, the existence of the soul, and life after death—PROVIDED those situations DID NOT include any telepathic possibility.

WHY this is the case is but a small tip of a gigantic iceberg.

TELEPATHY-THE PREEMINENT PENETRATION MODALITY

Whatever may or may not be said regarding telepathy, two very clear and unambiguous statements can be made about it.

First, it CAN be said that IF telepathy exists, then it would be of such overreaching and extraordinary importance that all Earthside institutions would have to be "reorganized" in the face of it.

Second, if a wide enough overview is accumulated about telepathy, and about how it is generally treated by Earthsiders, it is one human faculty that has a most excellent chance of being summarily shot down before it has a chance to open and wink its all-seeing eye.

The most visible explanation for this is that telepathy penetrates MINDS—and so its development is definitely cast into troubled waters where any format or element of mental secrecy might be involved.

It must be more or less admitted that most Earthside human activities cannot really get anywhere unless they are mounted upon this or that format of motivational secrecy or hidden agendas.

Using this situation as a simple rule of thumb, one can then easily grasp the extent and nature of the anti-telepathic antipathies that can be generated and exerted from the top of societal pyramids down into the populations beneath them.

I have no hesitation in stating the above, because a full part of it is a fallout based on real experiences of mine. As but one significant example, for fifteen years I was involved in secret developmental Psi work at the prestigious Stanford Research Institute. The work (in developing remote viewing) was largely funded by the U.S. intelligence agencies.

Because of this, many Washington types and many noted scientists visited SRI. Very many of them met only with my colleagues, and refused to meet little Moi, so much so that they would not even take lunch with me.

The reason: "Jesus, he can read my mind! I can't let him get anywhere near me." This quote is NOT paraphrased.

One of the amusing aspects of this is that IF telepathy is what it is, then one need not be in the proximity of a telepath in order to have their mind penetrated.

Another amusing aspect is that the funding agencies did sponsor the secret developmental work in remote viewing-somewhat on the grounds that it penetrates things, not minds.

This is to say that remote viewing pertains to penetration of "physicals," not to penetration of "mentals."

In any event, the principal reason why ALL formats of Psi research are marginalized, treated to energetic diminishment, or suppressed altogether is that those formats do include potentials too near the hated and unwanted telepathic faculties.

So, the whole barn of psychic research must be burnt down as quickly as possible, making sure that the telepathic horses don't escape.

There is one notable exception to this, and one utilized for creative cover-up purposes. This exception involves the discovery of approaches to telepathy most noted either for the fact that they DO NOT work, or because they serve to disorient and defeat approaches that MIGHT work.

Thus, the concept that telepathy is a mind-to-mind thing involving a sender and a receiver has been given extraordinary publicity—and has in fact become the principal Earth-side cultural model for it.

Intellectual phase-locking into this non-productive model is so intense and so widespread that Earthsiders literally cannot think of telepathy in any other way.

With the exception of some few experiments in the former Soviet Union, and in the Peoples Republic of China, the sender-receiver model has not yielded anything more than slightly above-chance results.

Even if the slightly above chance statistics are jerked around a little, none of them approach anything like telepathy plus.

And yet the sender-receiver model of telepathy has been clung to for a little over a hundred years.

As my own information package about telepathy increased, it was logical enough to first assume that since telepathy could be seen as a threat to all sorts of Earthside secrecy factors, those same secrecy factors would not, with any sense of humor, look upon the development of truly effective penetrative types of telepathy.

This probability still remains paramount, and clearly has an Earthside basis that can easily be established as such.

But if one approaches the concept that extraterrestrial intelligences might indeed be in possession of telepathy plus, then the Earthside picture that seems so certain all by itself, can easily take on some larger and astonishing dimensions.

Earthsiders can think that if Spaceside entities exist, then they are possessed of intelligence, and the same Earthsiders can indeed assume that alien intelligence to be, as often stated, "superior" to human intelligence.

After all, the Spaceside entities can build craft exceeding the limits of Earthside scientific knowledge. And so not only their technology, but their "minds" as well MUST be superior.

Even so, the only mind-models Earthsiders have for "mind" are their own rather limited versions of what mind consists of and from this Earthside model has been sanitized all factors that Earthsiders themselves don't want to consider or put up with.

Thus, Earthsiders project THEIR minds as conceptualized upon all potential extraterrestrial entities.

In this sense, the intellectual phase-locking regarding mind is plant wide, with the final situation being that the further one moves upward in Earthside power structures the more constricted that phase-locking becomes.

Thus, there is some pungent and meaningful kind of hidden story here. But whatever it is, it clearly begins with the fact that Earthside science, philosophy, religion, sociology and psychology DO NOT sponsor research into what can collectively be called Psi, while those same noble institutions are rather noted for condemning it.

The modern Space Age facilities need not bother with the existence of extraterrestrial minds because those same facilities insist that nothing of the kind exists, near Earth anyway.

Most surprisingly, one might think that Ufologists would consider mental processes of extraterrestrials, since they are so energetically involved with extraterrestrial equipment and technology.

None of the above will touch the topic of Psi with a ten-foot pole, and all of the above protest any feasible, positive necessity for acting any other way, although some psychologists studying abduction phenomena have begun to notice the telepathic factor.

At least two observations can be made relevant to the above. First, one might consider that the Earthside retreat from Psi is something akin to protesting too much.

Second, if I were an ET with highly developed Psi skills (and which might have led in the first place to the evolution of superior technology), I wouldn't particularly want Earthsiders to develop Psi faculties.

And if telepathy was an element in, say, consciousness universal, I'd soon figure out how to telepathically impregnate Earthside human consciousness with intellectual phase-locking that was detrimental to positive telepathic plus development.

The reason might be very obvious. After all, what ET would want Earthside telepaths penetrating Spaceside affairs, especially, perhaps on the Moon so near to them?

Thus, in this, at least, Spacesiders and Earthsiders might have something in common—the Telepathy War, won hands down so far by the Spacesiders.

THE EARTHSIDE CONCEPT OF TELEPATHY

In the previous chapter, I pointed up that the modern concept of telepathy has not produced much in the way of evidence for telepathy much above some very low threshold activity.

In other words, human telepathic faculties are known to exist. But, with the exception of spontaneous examples of telepathy, it does not function in a high-stage way.

There could be any number of reasons for this. But one reason is that the concept is at odds with what telepathy really consists of. Because that concept is assumed to be so correct, it is never questioned—resulting in failure to move beyond it.

This is the same as saying that the concept is so widespread that strong intellectual phase-locking of the concept has taken place.

During modern scientific times, the standard images of telepathy usually picture two heads or brains facing each other. The two heads or brains are meant to represent two MINDS. But since no one seems to have figured out how to render a mind into a pictorial image, an image of a head or brain stands in for one.

Between the two heads or brains are usually placed something like squiggly lines.

The squiggles are meant to be suggestive of vibrations or waves telepathically traveling from one mind to the other mind. Sometimes one of the two heads is indicated as "sender," the other as "receiver." Since telepathy is identified with thoughts, the squiggly lines are meant to represent them.

The modern idea fundamentally holds that telepathy is MIND-TO-MIND, and that the brain, or at least the head, is assumed to be the Seat of the mind or the mind itself. This fundamental idea seems entirely logical.

However, the above only represents the chief THEORETICAL

model of telepathy as envisioned by some early psychical researchers about a hundred years ago.

But because of its apparent logic, the theory was assumed to be the truth of telepathy.

Since the theory seemed so logical, the mind-to-mind concept quickly underwent wide-spread intellectual phase locking to the degree that it soon obtained the planetary-wide status of unquestioned and unchallenged consensus reality.

Whether things are true or not, consensus reality usually casts them into cement. Thereafter, it is very difficult to tamper with a consensus reality—especially one that has "gone planetary," so to speak.

But if the modern concept of telepathy is somewhat dispassionately examined, then, as we will shortly see, the first and major problem encountered relates to where and to what the mind is—and to IF it is.

Beyond that, we can see that the modern concept of telepathy has hardly any long-term historical tradition which would establish it as a natural constituent of our species.

So one has to rummage around in early history in a kind of archaeological dig in order to discover what there was in the way of antecedents to telepathy.

The ancient Romans identified two major terms which apparently referred to two different kinds of thought processes. We continue to use them today, but in quite different ways.

The Latin INTELLECTUS referred to the processes of thinking while in the awake state. The thinking was based on the physical senses, but included the senses of emotional feeling, the will, and decision-making based on perceived evidence.

The Latin INTUITUS was taken to refer to anything that did not fit into the parameters of INTELLECTUS, but which anyway influenced persons AND what happened or was to happen to them.

It was considered that INTUITUS was greater than individuals, but that individuals had a kind of intuitive thought processing capability. Some had more of this INTUITUS than others, and so INTUITUS was a Roman extension of the great traditions in antiquity regarding shamans, oracles and seers.

This great tradition was world-wide, and far antedated even the ancient Romans and Greeks. In this very ancient tradition, it is quite probable that what we now specify as clairvoyance, intuition and telepathy were all housed within the same concept and not identified separately.

The usefulness of INTUITUS was that it provided information to users, and they didn't much care how it was gotten, only that it was.

We have only to add our contemporary idea of mentally processing information to the concepts of INTELLECTUS and INTUITUS, and we come up with a rather clear picture of the past.

But like the ancients, we would have to specify different mental processes for different kinds of information. We do this by indicating that there is a difference between:

1. information derived from immediately objective sources; and,
2. information subjectively derived from sources that are not immediately objective.

The only real difference between the ancient and the modern ideas of intellect and intuition is that we today think of them as THINGS—while the ancients considered them as information acquisition processes or functions.

But there is one more quite subtle difference. When we think of intellect and intuition as things, we will then try to use our things as tools to acquire information. In this sense, we first position the tool ahead of the information it is supposed to deal with.

Since we think of intellect and intuition as things, we suppose that the ancients did likewise.

But the evidence is quite strong that they first positioned the information to be acquired by whatever means and didn't really conceptualize thing-like tools needed to acquire it.

This subtle tradition still goes on, albeit outside of modern science and psychology. Many highly functioning people want information—and they still don't particularly care how they get it as long as they do get it.

We well understand that between intellect and intuition quite different thought processes are involved.

However, since in our modern times we don't know what intuitive thought processing consists of, we attempt to utilize intellectual thought processing to achieve intuitive results.

The results achieved by this mismatching are not much better than chance expectation.

It was not until the sixteenth century that the concept of clairvoyance made its appearance in France. This commenced the distinction of separate INTUITUS factors.

In French, the term was first utilized in the contexts of keen insight, clearness of insight, insight into things beyond the range of ordinary perception. These French definitions are approximate to the early Roman idea of INTUITUS. The emphasis, of course, was on INSIGHT.

The route of the French CLAIRVOYANCE into English is not clear, but it seems it was not adapted into English usage until about 1847. When it did appear in English, it carried a slightly different definition: a supposed faculty of some persons consisting of the mental perception of objects at a distance or concealed from sight.

Unless the difference is pointed up, it probably won't be noticed. There is a strategic difference between the concept of insight and the concept of perceiving objects at a distance or concealed from sight.

Within the context of this book, the definitions of INSIGHT are somewhat amusing: the power or act of seeing into a situation; penetration; the act of apprehending and penetrating into the inner nature of things or seeing intuitively.

The use, in English, of the term CLAIRVOYANCE served to detach it from insight, and then to establish a special category limited to the "seeing" THINGS.

The emphasis thus shifted toward a specialty interest only as regards mental mechanisms via which clairvoyance might function.

With the English concept of clairvoyance thus established as seeing THINGS (not seeing insight, as it were,) it then becomes obvious that a companion category having to do with penetrating minds was necessary. After all, human experiencing IS involved with things AND mental activities.

This special category already existed when the English concept of clairvoyance came into existence.

The category was called THOUGHT-READING, and had a history going back for some centuries. The history was rather wobbly, though, since thought-reading had been used as a form of entertainment and was thus heavily occupied by frauds.

The only concept of real thought-reading that has survived down until today is expressed as someone "reading" someone else's "beads"—thereby gaining insight, etc.

In any event, the parameters of what might constitute thought-reading were vague—and also carried the disadvantage of being related to the idea that thought-reading could be "picked up" in group kinds of ways.

Such spreading about could, by some unknown subliminal means, result in infectious hysteria of what was latter termed "mob psychology"

What was wanted in order to break away from thought reading was a concept that specifically identified "direct action of one mind on another, independent of the ordinary senses." No such restrictions could be applied to thought-reading because of its somewhat notorious groupthink characteristics.

To fulfill the idea of direct action of one mind on another, the concept of THOUGHT-TRANSFERENCE appeared in England between 1876 and 1881.

However, this concept was short-lived, because it remained somewhat cluttered with a confusion revolving around the idea that some kind of trance-like rapport was involved regarding the transference of thoughts and emotions. The transfer of emotions was still quite close to group responsiveness via some kind of entrainment.

All of these problems were gotten around (or so it was thought) with the coining, in about 1882, of the term TELEPATHY by the psychical researcher, F. W H. Myers.

One of the most cogent summaries of telepathy is found in the 1920 *Encyclopedia of the Occult* compiled by Lewis Spence.

Therein we read that "The idea of inter-communication between brain and brain, by other means than that of the ordinary sense-channels, is a theory deserving of the most careful consideration."

Compacted this way for research purposes, "The idea" sounds absolutely great, doesn't it?

Well, as already mentioned, "The idea" represents the chief horror of all horrors—in that very few humans relish the idea of having their brains penetrated in this way at all.

As Lewis Spence (among other of his contemporaries) noted in 1920, inter-communication by means other than that of the ordinary sense-channels is something deserving of careful consideration.

But this implies that there would have to be a desire to commence the consideration in the first place. After all, one has to establish the need or willingness to consider something before one can go ahead and "carefully" consider it.

Since the idea of telepathy is somewhat in conflict with preserving the idea of secretive power, the road of telepathy begun in 1882 was to find itself filled with major social blockages.

In any event, Myers established a rather precise definition for the new term: "a coincidence between two person's thoughts which requires a causal explanation."

The "causal explanation" was theorized as being like radio broadcasting "waves" which were sent and duplicated by receivers known as radios.

TELEPATHY replaced the earlier term, THOUGHT TRANSFERENCE, largely because the latter did not avail itself of the radio-wave hypothesis as THE causal explanation. Thereafter, the image of telepathy I've outlined at the head of this chapter has held complete sway.

However, and as established, since telepathy cannot really exist without its major substance—thoughts-the telepathy situation still revolves around thoughts and their direct transfer from one brain to another.

We now encounter the first of the major stumbling blocks. Everyone realizes that a thought contains information. And so here we are in the vicinity of a quite good analogy—a bottle of wine. Thoughts are the wine. But what does the bottle consist of?

THOUGHT is one of those terms that have many definitions—

too many to bring any clarity to the issue.

THOUGHT: the action or process of thinking; serious consideration; recollection; reasoning power; the power to imagine; something that is thought; the individual act or process of thinking; intention; plan; the intellectual product of organized views and principles of a period, place, group, or individual; characterized by careful reasoned thinking.

As an addendum to the above definitions of THOUGHT, some, but not all, dictionaries also attach the term MINDFUL, the principle definition of which is "inclined to be aware."

So, one can read all of the eleven definitions of THOUGHT— and observe that thought-activity of any or all of them COULD proceed without any professed inclination to be aware of anything.

In the event of this, however, only the grossest cases might become noticeable. They would be dubbed as MINDLESS—that term referring to "inattentive, destitute of awareness, mind, or consciousness."

All of the above might seem like extraneous excursions into words. But actually, one might well wonder if someone would telepathically pick up someone else's mindless thoughts—such as utilizing rather mindless and dull cards of symbols and color shapes to test for telepathy.

As it was back in the nineteenth century, most of these definitions for THOUGHT, and the confusions they carried, were easily available. And so Why-O-Why that term was seized upon at all as relative to telepathy is virtually inexplicable.

A vastly more cogent term would have been INFORMATION TRANSFERENCE.

As to TELEPATHY, this was a neologism put together from two terms: TELE meaning across; and EMPATHY traditionally referring NOT to thought, but to "the capacity for participating in another's feelings or ideas as a result of becoming infused with them."

INFUSE is taken to mean to pour in, to introduce into, to insinuate, inspire, and to animate.

If the reader has found all of the above to be more than a little confusing, well, don't worry.

The concept of telepathy makes perfectly logical sense—IF it is discussed WITHOUT including its attendant difficulties.

If the attendant difficulties are mentioned, then various cognitive problems begin to arise—largely because the assumed logic of the telepathy model DOES NOT consider the "bottle" that contains the wine (thoughts).

EARTHSIDE GROUPTHINK

I have outlined the theoretical concept of telepathy as a mind-to-mind thing made possible by something akin to radio broadcasting waves. I have also pointed up that that model is universally accepted even today as THE correct and only model of telepathy.

And I have more or less challenged the authenticity of that model—largely because nothing has ever developed out of it. Yet, in spite of its demonstrated unworkability, the model is stubbornly clung to by Earthside groupthink on a world-wide basis.

The reader might assume that my challenge to its authenticity originates with me. But this is not the case at all.

The concept of telepathy as mind-to-mind came into existence in 1882, and was quite inspirational. Because of what was involved, it was given a very thorough working over during the following twenty-five years.

On the one hand, no real advances were achieved, while on the other hand evidence mounted indicating the theory was neither correct nor applicable.

The latter situation was summarized in 1919 by James Henry Hyslop, a former Professor of Logic and Ethics at Columbia University, and one of the most distinguished American psychical researchers.

Hyslop published a lengthy review of previous telepathy research and ended up with a six-part statement that "There is no scientific evidence for any of the following conceptions of it."

∅ Telepathy as a process of selecting from the contents of the subconscious of any person in the presence of the percipient.

∅ Telepathy as a process of selecting from the contents of the mind of some distant person by the percipient and constructing these acquired facts into a complete simulation of the given personality.

Ø Telepathy as a process of selecting memories from any living people to impersonate the dead.

Ø Telepathy as implying the transmission of the thoughts of all living people to all others individually, with the selection of the necessary facts for impersonation from one individual by another individual.

Ø Telepathy as involving a direct process between agent and percipient.

Ø Telepathy as explanatory in any sense whatever, implying [involving] any known cause.

Thus, the news that the theory of telepathy didn't work was available in 1919. Why this evidence was trashed, and why Earthside groupthink continued to advocate the unworkable telepathic theory is a question that few have ever considered.

It is perhaps unfortunate that Professor Hyslop published the six findings listed above in his 1919 book entitled *Contact With The Other World.* This topic indeed placed him outside of science and philosophy proper.

As it was, continuing confidence in the unworkable telepathy model was so high that its enthusiasts have simply proceeded advocating it down until the present.

One hypothetical answer as to why Earthside groupthink has continued to be infected with the unworkable model of telepathy is that it DOES NOT work.

As long as Earthsiders are intellectually phase-locked into the assumption that an unworkable model is none the less authentic, well, Earthside secrets will remain unpenetrated by telepathic modalities.

If this is the case, then it is not the failed model of telepathy that is important, but rather the Earthside groupthink that promulgates acceptance of its authenticity.

Here, then, is a recognizable case of information package

management to defeat the development of Earthside telepathy. It does so merely by instituting an information package from which telepathy doesn't stand a chance of being developed.

THIS is like placing and reinforcing a mental screen seemingly so logical that it obscures its own illogic.

We can see that this type of screen would be thought of as advantageous to Earthsiders who would not be thrilled if their secretive activities were to be telepathically penetrated.

One might hypothetically also consider that the ET might likewise NOT be thrilled for much the same Earthside reasons. Thus, Earthside telepathy may be doubly damned, hypothetically speaking of course.

In any event, the nature of Earthside groupthink is quite interesting, in that really effective management of information packages can take place only if groupthink truly exists and that it does have some kind of telepathic basis.

Otherwise, attempts to manage information packages on an individual basis would be quite laborious.

Groupthink is acknowledged as existing. It can be seen in the way Corporations seek to "condition" their employees on behalf of being enthusiastic about the Corporate structure and its goals.

Additionally, the concept of groupthink and the concept of intellectual phase-locking seem to have something to do with each other.

Both groupthink and intellectual phase-locking appear to be extensions of the age-old axiom that birds of a feather flock together—while THIS is assumed to be an active element of human nature.

If the existence of groupthink and intellectual phase-locking is accepted, then the only remaining problem, or opportunity, is what information packages are to be inserted into them and thereafter managed for one end or another.

However, in the light of the above the existence of group-minds cannot be escaped.

And if anyone wants to discover one single topic that is constantly bleeped, avoided, and suppressed, you only have to consider the nearly complete absence of this one.

To get into this, even if only partially, it is necessary to start by considering the nature of what Earthsiders have elected to identify as consciousness.

IS CONSCIOUSNESS INDIVIDUAL OR UNIVERSAL?

I f one begins to examine the Earthside secrecy regarding UFOs that is by now apparent almost on a daily basis, it is appropriate to first focus on exactly what is being kept secret.

If one thereafter progresses beyond the obvious, one soon finds that not only is information being kept secret, but that dis-information is being supplied from very high levels to disable and cover up information that can't be kept secret.

Thus, the UFO situation is characterized by secrecy barriers and by cover-up stratagems.

There are two factors about this double situation that are remarkable, but which seldom are commented upon.

The first factor has to do with the social dimensions involved. It is quite fair to say that the dimensions are worldwide, or, put another way, planetary.

This factor leads into the second one, the fact that an enormous cooperation is required to keep the secrecy and the cover-up in place through the decades in which both have been implemented and maintained.

The whole of this, of course, is something of a charade in that UFOs have been seen, photographed and video-taped all along. Thus, as this chapter is being written in 1998, the general public dwelling in most nations thoroughly realize that UFOs exist, and that they are operated by intelligence.

If one meditates on all of the above, it can become somewhat clear that the existence of the UFOs is not what is being covered up—because they ARE seen, photographed and videotaped.

Additionally, the idea that the craft are the products of an intelligence can't be covered up—largely because the idea that they are NOT the product of an intelligence is ludicrous.

After wending one's way through the mysteries involved, one can be left with the rather stunning question: WHAT IS IT that is actually being clothed in secrecy and cover-up?

After all, the UFOs are visible planet-wide (and actually on a daily basis if one reads the weekly UFO Update now available via the Internet.) Additionally, the secrecy and the cover-up are trenchantly visible, for they have been adequately exposed in a great number of books.

In the light of this, about the only place the secrecy and cover-ups are being effective is among those responsible for both. This is to say, among government, military, scientific and media hierarchies—all of which remain quite mum about whatever it is those authoritarian structures are remaining mum about.

And whatever this IS, it is not clear at all.

To emphasize: Covering up the obvious is an oxymoronic exercise. But covering up something ABOUT the obvious that is not readily apparent via the obvious evidence could make sense out of what is otherwise only a silly charade.

Every aspiring intelligence analyst proposing to work within secret agencies learns that one way to break a mystery that won't yield to easy explanation is to look around for mysteries that are somehow similar.

In this case, the secrecy and cover-ups are being maintained, rather Big Time, by government, military, science and media collaboration. Therefore, it is useful to look around for another example which those Big Four entities ALSO collaborated in covering up.

One example along these lines comes to mind. This involves an issue that is a little difficult to articulate because it is as energetically suppressed and covered-up as is the issue of ET visitations and intelligence.

A tip of this particular iceberg first surfaced in 1957 when the writer Vance Packard published a book entitled *Hidden Persuaders*. The original meat for Packard's book is given as follows.

In the early 1950s, the owner of a movie theater in New Jersey had apparently learned something about subliminal suggestion. He contrived to briefly flash the words "Drink Coca-Cola" over Kim Novak's face. This resulted in a 58 per cent increase in Coca-Cola sales over a six-week period.

Packard's *Hidden Persuaders* gave depth and substance to this phenomenon and described how large groups of human minds could be influenced by words or images flashed so quickly that the intellect could not perceive them, but that the subconscious did. Indeed, the fact of subliminal communication and perception was obvious.

Even so, the resulting brouhaha was absolutely enormous, and the Big Four cooperated in establishing negative information packages the purpose of which was to condition public awareness away from the reality of subliminal activities.

If the conditioning steps are examined, it can be seen that they were not entirely unlike those being promulgated regarding the UFO cover-up situation. I.e., to deny, discredit, and decrease confidence.

There are several ways to assess the Vance Packard situation. Eldon Taylor examined it in his book *Subliminal Communication* (1988). As Taylor wrote: "Packard presented a case for persuasion through the art and science of motivational analysis, feedback, and psychological manipulation.

"*Hidden Persuaders* was the first open attempt to inform the general public of a potentially Orwellian means to enslave the mind and to do so surreptitiously."

It would have seemed that Packard's book could have been taken culturally in stride since it was no secret (1) that minds could be influenced, and (2) that they were influenced by art, literature, intellectual suggestions, and educational conditioning.

After all, the major goal of any social grouping is to achieve broad intellectual phase-locking, so as to benefit from melded group-mind responses and thereby maintain the contours and workability of the society.

As it was, the Big Four carried on in ways that amounted to a rampage against subliminal perception—and the issue was thereby slowly re-submerged beneath the awareness of public cognizance.

Big Four outrage surfaced again in the early 1970s when yet another book appeared entitled *Subliminal Seduction,* authored by one Wilson Bryan Key. This book quickly underwent several printings by various publishing houses.

So an extra-large dose of negative deconditioning response emerged from the Big Four. The general tenor of the Big Four deconditioning responses verged on apoplexy which may have induced much the same in the public mind—i.e., sudden diminution or loss of consciousness, sensation, and voluntary brain motion.

Even so, the Big Four reactions were so large that many began to suspect that where there was so much cover-up smoke that there must be a goodly fire. And so the book became much in demand.

Key's book provided substantial evidence that subliminal seduction was being utilized by big-time Madison Avenue advertisers in a conscious effort to influence the public mind in order to increase sales of various products through the integration of hidden messages.

For example, it had been learned that embedding subliminal "messages" in ad illustrations by way of very subtle images of naked women or the words FUCK, SUCK, TITS or BALLS, indeed increased sales of what was being advertised.

The subtle embeds do not work with regard to conscious perception, but rather stimulate activity in the subconscious level where drives or urges for something originate. This results in perception without awareness.

It was ultimately confirmed that subliminal "messages" could induce activation or deadened public responses to just about any issue.

In any event, the so-called "controversy" went bananas. It was summed up in a very hefty and scientifically respectable book entitled *Subliminal Perception: The Nature of a Controversy (1971),* authored by Norman F. Dixon, then at the University College, London. Dixon's book was never published in the USA as far as I know.

Aside from the elite's obvious efficiency in managing the "public mind" this way or that, the issue of subliminal seduction is clearly attached to the issue of the group-mind. For the "public mind" is, after all a group kind of mind.

The public mind, as a group kind of mind, also is referred to as mass consciousness or mob consciousness.

If one then expends the time and effort to troll for information about mass consciousness, one will encounter a very strange factor regarding cover-ups of information packages.

This must be preceded by mentioning the obvious desire of public managers to understand "human behavior" and how mass human consciousness functions—in order to better mind-control the public mind this way or that.

It is thus unthinkable that no research along these lines has ever been undertaken.

My own research into this area revealed that mass consciousness or mob consciousness research came to an abrupt end in about 1933-1935. This is to say, that it came to an end as far as public access to it is considered.

It ended because of a set of discovered conclusions. Among them, that mob consciousness responded collectively NOT to rational intellectual perspectives, but to some kind of emotional empathy that was somehow subconsciously TRANSMITTED. This, however, could not be explained unless the concept of telepathy was brought into consideration.

And THAT was the end of THAT kind of research. But here is a rather remarkable link of some kind. If the existence of developed telepathy is put down and covered up by elitist Earthside forces, then if there might be a telepathic Spaceside connection, the existence of that particular factoid would need to be covered up.

It is worth repeating that psychical and parapsychological research more or less bit the dust BECAUSE it proposed to research telepathy—the one human attribute that many Earthside power structures prefer NOT to be developed.

However, in order to get just a bit deeper into this possible issue, it needs to be approached from a slightly different angle. This involves the matter of consciousness.

There are so very many definitions of CONSCIOUSNESS that they altogether assume the guise of a cognitive sump. But even so, there is an official definition of it, and it is this one that the Big Four more or less cling to.

This definition, in its several parts, is found in *The Encyclopedia of Philosophy,* published in 1967. The definition is not obsolete, however, since it remains more or less in force today.

The definition begins with a reference to John Locke (1632-1704), the renowned English philosopher and founder of British empiricism.

Locke defined CONSCIOUSNESS as "the perception of what passes in a man's own mind...[as the process] of a person's observing or noticing the internal operations of his mind. It is by means of consciousness that a person acquires the ideas of the various operations or mental states, such as the ideas of perceiving, thinking, doubting, reasoning, knowing, and willing and learns of his own mental states at any given time."

The Encyclopedia then goes on to clarify that although the term CONSCIOUSNESS has many definitions, it "has a broad use to designate any mental state or whatever it is about a state which makes it mental.... It is consciousness which makes a fact a mental fact."

Considering the many ambiguous and confusing definitions of CONSCIOUSNESS, the above offers a clinical efficiency that can hardly be doubted. Thus, most would take it at its apparent, and important, face value.

But the definition establishes a parameter that is quite interesting, once it is pointed up.

For the definition consigns the definition of CONSCIOUS to existing within the mechanisms of the individual. This is to say, that although each person has consciousness, it is none the less individual to that person. For increased clarity, each person has consciousness, and thus each is, so to speak, an island of consciousness among multitudes of other islands of the same.

If, then, information is transferred between the islands, it has to be accomplished by objective means.

Nowhere in the Encyclopedia entry is there any hint that consciousness is anything other than individual.

Thus, but without saying so, telepathy as the melding of consciousness independent of objective means of transfer is forbidden.

There is no entry for TELEPATHY in the Encyclopedia. But there is a rather fair synopsis of ESP PHENOMENA, in which telepathy is referred to as a "species of ESP," but within which nothing is learned about it—except an admission that it exists.

As it is, telepathy cannot exist, much less be explained, IF the parameters of consciousness are limited to the mental equipment of the biological individual.

Since information is "exchanged" or "acquired" between human individuals in the absence of any objective methods to do so, and in that the information so exchanged results in mental perception of it, it is obvious that a format of consciousness exists that is independent of each biological human unit.

The Encyclopedia definition thus seems good as far as it goes but is nonetheless incomplete.

And that definition has deficiencies. For example, it stipulates that consciousness is mental awareness. But long before the Encyclopedia was compiled in 1967, the real existence of the subconscious was confirmed. The principal definition of the SUBCONSCIOUS holds that it is aware of information that the mental awareness is not aware of.

Not only that, but that the subconscious causes the bio-mental organism to RESPOND to information that the mental awareness is not aware of. And indeed, THIS is the working hypothesis that leads to the efficiency of subliminal "messages."

Additionally, the early mob consciousness research resulted in the considered estimation that information WAS transferred and exchanged at some emotional sub-mental-awareness level.

As a result, some kind of sub-mental union or bonding resulted in what could only be thought of as an unknown kind of telepathy that served to induce behavior of a group-mind force.

One of the concepts that can come out of this is that although each individual may be an island of consciousness, all such islands might be residing in a greater ocean of consciousness which exists independently of each human life unit.

In this regard, the Encyclopedia definition establishes that consciousness IS only what the individual becomes mentally aware of.

But strictly speaking, the definition is describing a FUNCTION of consciousness, not, so to speak, the "substance" of consciousness itself.

And with this, we could now plunge into the intricacies of mysticism whose chief proponents have always held that consciousness is a universal substance, and that each human is only a small manifestation within it.

But I'll shift direction here, in order to get back to the point of this chapter, and indeed this book.

If Spaceside extraterrestrials do exist, and there is plenty of Earthside evidence of them, then one has to wonder about THEIR consciousness. For example, is their consciousness the same universal stuff of human consciousness?

We might also have to wonder if THEIR consciousness is more "technically advanced"—say, something along the lines of their "advanced material technology" so advanced, indeed, that their craft easily disobey the known laws of Newtonian, atomic and quantum knowledge on Earthside.

We might even be inspired to wonder if, in their advanced consciousness technologies, they would remain as klutzy as Earthsiders regarding ESP and telepathy.

We might also have to wonder if their telepathy is a developed version of a telepathic "language" that is universal within universal consciousness. Others before me have indicated that if consciousness exists, then it must have operative "laws," and cannot possibly consist only of what a given, individual bio-mind Earthside entity becomes mentally aware of.

If this consideration is given enough extrapolation, however, it could increase the possibility that it might be to someone's benefit to utilize the laws of advanced consciousness technology to ensure:

1. that Earthside entities DO NOT become mentally aware of a lot of things, and
2. that Earthside entities DO become conditioned to be mentally aware only of what someone wants them to be aware of.

The above two possibilities are only very speculative, of course. But if such Earthside mental management was indeed factual, then any ostensible success would depend on DELETING (or at least confusing) certain factors from human mental awareness.

There may be many of such needed deletions. If I wanted to accomplish (1) and (2) above, I'd delete concepts of consciousness that extend beyond individual functioning.

I'd also delete, or at least suppress, the Earthside discovery and efficient applications of subliminal messages and suggestions.

After all, their relevant techniques are effective toward group-mind management and groupthink parameters—and especially with regard to which information packages should or should not be intellectually phase-locked upon.

It would also be useful to ensure that different groups of Earthsiders intellectually phase-lock on different and contrasting information packages. This would not only keep the groups confused by each other but might even keep them antagonistic. And so the concept of Divide and Rule would then be a piece of cake.

All Earthside efforts toward discovering and developing ANY kind of telepathy would have to be vigorously stunted from the get go—because if Earthside telepathy can penetrate Earthside minds, then there is no reason why Spaceside "minds" cannot be penetrated as well.

Having established such goals, I'd then have to figure out how to implement them Earthside, while at the same time ensuring that the goals being implemented Earthside remain thickly covered up.

Fortunately, in this regard, Earthsiders intellectually phase-lock quite easily, often in a massive way.

If I were a Spacesider doing this, I would have access to telepathy plus. And so all that would really be needed are a few subtle tele-powered messages that enter subliminally into the rather backward order of undeveloped Earthside consciousness.

As two additional blessings for Spacesiders, Earthsider elites are usually intellectually phase-locked on the thrill of having secrets, and so they keep everything as secret as they can. This automatically leads to the necessity of covering up their secrets.

And so, as a general prophylactic measure, they usually cover up everything they can.

The foregoing is, of course, a foray into gross speculation—and has, as it does, many holes in it.

But back in Earthside realities, there runs one consistent theme throughout. This is the perpetuating disenfranchisement of telepathy and all that its penetrating aspects it might imply.

POSTSCRIPT

LOTS OF WATER ON THE MOON

During 1998, while this book was being produced, two major scientific developments were announced concerning the "discovery" of water and atmosphere on the Moon.

It is important to mention these because they are certainly relevant to this book, and because in some quarters they have again aroused enthusiasm about the possibility of colonizing the lunar satellite.

These recent developments are momentous and wonderful, to be sure. The Moon is no longer the dead, arid, airless and uninhabitable satellite that ALL official sources since the 1920s have insisted it was.

Not only is the Dead Moon Dictum now almost magically and abruptly overturned, but these two lunar developments make it seem as though official science is marching onward in some kind of full-disclosure fashion.

However, one must keep in mind that this is the same Moon that was the expensive colonizing target of the American and Soviet 1960s Space Age efforts, the same Moon that was frequently orbited, upon which men walked, and the same Moon no one went back to.

And if one knows something of the Moon's many shocking oddities and anomalies, it is clear that there are numerous lunar factors still lingering in the cover-up scenarios.

As we have seen, those factors are not insignificant. Collectively accumulated by numerous unofficial observers utilizing official documents, evidence for them is copious, direct and quite compelling.

As to the water, it is said to be in the form of ice, mostly at the poles and buried about half a meter beneath the lunar surface.

The estimates are impressive regarding how much of it there is: some six billion metric tons. This is said to be enough to sustain upwards of 100,000 lunar colonists for a century and also provide a fuel source of oxygen and hydrogen for Moonbases and space travel.

While this is exciting news, if the evidence is taken into account for earlier-known lunar clouds and mists clearly visible in some officially released photos of the 1960s, then one cannot think that ALL of the lunar water is only in the form of subsurface ice.

As any dictionary or encyclopedia will confirm, a cloud is defined as "a visible mass of particles of water in the form of fog, mist, or haze suspended at some height in air or atmosphere." Thus, if the Moon did not have an atmosphere, the mass of water particles would have nothing in which to suspend.

As to the lunar atmosphere, the American Geophysical Union recently indicated that although "conventional wisdom says the Moon is devoid of atmosphere, and in layman's terms this may be close enough to the truth, the space just above the lunar surface is not a total vacuum." (See: AGU Release No. 98-26,17 Aug 98).

There is, of course, no doubt that the lunar atmosphere is not like Earth's. But even if more tenuous and not as thick, the lunar atmosphere now OFFICIALLY exists, as does the lunar water.

Thus, the UNOFFICIAL sources of the past that referred to the existence of lunar water and atmosphere have turned out not only to be correct, but ironically vindicate their authors.

One of the sardonic fallouts of this is that the materials, including official NASA photography, published by the unofficial sources (see bibliography) might be read with renewed interest.

Official NASA photos that clearly show lunar clouds and mists have been available all along dating especially from the days of the Lunar Orbiters and manned Apollo craft.

The presence of clouds and mists is an undeniable indicator of available water vapor and atmosphere. So one can wonder WHY their existence was unequivocally denied by officialdom in the direct face of the available photos.

Only the amounts of the water and atmosphere would have been in question. Yet the official stance held that there was none of either.

One can hypothetically deduce, as almost all eagle-eyed unofficial analysts did, that the lunar water and atmosphere cover-up was not in the name of science.

It obviously involved other factors which, themselves, must have had some kind of strategic importance regarding why a cover-up should exist in the first place and then be maintained for over sixty years.

Indeed, if one thinks this through, there was no NEED during the 1960s Space Age to cover-up water and atmosphere since these would have added a great deal to the enthusiasm to colonize the Moon.

Just beneath the surface of the irony, though, are a number of factors that probably will be smoothed over, if not completely erased from lunar history.

The evidence for lunar water was scientifically noted and written about by the early selenographers of the latter part of the nineteenth century and early decades of the twentieth. Later, the analyses of the selenographers were confirmed by official NASA photographs.

To emphasize, as most of the unofficial analysts pointed up, lunar clouds and mists drooping over crater rims can be seen in numerous official NASA photographs that were achieved during the

1960s.

And so it is to be rapidly conceded (or should have been at least) that where there are clouds, then water and air are not far off. After all, it takes air and water to make clouds.

As but three examples of NASA photos showing clouds, the following have been published in several unofficial sources:

An unmistakable Mackerel type cloud bank can easily be identified just off the crater Vitello (NASA Lunar Orbiter V photo, No. MR 168).

A very large cloud bank hangs over the rim of a crater in Mare Moscovience on the Moon's backside (NASA Lunar Orbiter V photo, No. HR 1033).

However, the same photo shows what seem be two cigar shaped airborne objects casting shadows on the surface. The photo also shows a very large, circular dome of which the Moon is known to have many, some of which appear and disappear.

Two large cloud banks are seen hugging the edge of crater Lobachavsky (NASA Apollo 11 and NASS Apollo 16 photo, No. 16-758).

But most remarkable and quite clear in this photo is a large, undeniably round object poised near the top of the crater wall and casting a dark shadow down-slope.

This is not a dome that might be a natural formation. It is a round object, or structure, circular in all dimensions. It reminds one of a golf ball sitting neatly on a tee.

Whatever it is, it must be extremely large since it is clearly distinguishable in the rather low-resolution photograph. So one wonders if the higher-resolution lenses of the military Clementine craft zoomed onto THIS particular "structure." After all, it is said that Clementine "mapped" most of the Moon.

However, any high-resolution lunar evidence of any kind remains absent. Of course, detecting ice BENEATH the lunar surface does represent some kind of high-resolution capability. It therefore seems logical to think that what is ON the surface might be detected, too.

And so once again we are brought back to the conflicting nature of the official and unofficial versions of the Moon.

The OFFICIAL versions emanated from the combined auspices and gargantuan systems of government, science, academe, and major media. The official versions long held that there could not be water (or atmosphere) on the Moon. So all official reports to the public were geared to reinforce the idea of the absence of water.

The UNOFFICIAL versions emanated from numerous individuals some of whom obviously spent considerable research time, effort and money to produce their books and articles. Among these, for example, was Fred Steckling (see bibliography), whose 1981 book detailed the existence of lunar water (and much more, such as vegetation and artificial structures).

The unofficial versions were of course trashed by various and sometimes nefarious activities of officialdom.

Now that the existence of lunar water has been confirmed, the better of the unofficial sources are vindicated at least as far as water and atmosphere are involved.

However, if the unofficial eagle-eyes could detect evidence for lunar water as early as 1981, then it is almost certain that the same eyes can detect OTHER lunar factors, too, and draw appropriate conclusions about them.

For example, even official low-resolution photos acquired from NASA show many massive golf ball "things" in the most unlikely lunar places. These and the large domes that appear and disappear, are not as hard to detect in the official photos as is the water.

This author is not the first to notice that voyaging to the Moon abruptly ceased some twenty-five years ago and did so after the utterly enormous expenditures of getting there in the first place.

In attempting to identify official explanations for this "loss of interest in the Moon," the one most frequently encountered is (believe it or not) that THE AMERICAN PUBLIC had become disenchanted with costs and results of the NASA Moon program.

It is true that the American public SOMETIMES can influence major affairs. But the Soviet Union also stopped going to the Moon. In the former Soviet Union what the Soviet public thought about anything did NOT matter at all. Anyway, as it turned out on the American side, NASA stopped its expensive Moon colonizing goal but promptly undertook even more costly space age projects in different directions.

Thus, the Moon disappeared into anonymity behind all of the other space age projects even though science, NASA and government insiders certainly did know of the water and atmosphere potentials that made the Moon ultra-ripe for colonization.

It certainly takes a rather simplistic gullibility to accept that the American efforts to colonize the Moon were abruptly canceled because the public, of all things, had lost interest.

Indeed, during the 1960s twenty manned Apollo craft and launch equipment had been planned, each paid for at great cost, each built and relatively ready to go.

Yet, only seventeen Apollo missions lifted off, while the remaining three were abruptly terminated.

So, water and atmosphere on the Moon. We had landed there several times albeit in locations where nothing more than the soil and rock immediately beneath could be seen. No high-resolution photos of lunar vistas as seen from the lunar surface were ever released. Great footage, though, of the sand in which the Flag was planted, the one which inadvertently started flapping in the lunar breezes.

Buried in the cover-up are THREE unused Apollo crafts. It is perfectly logical to want to find out why they were left to rust and rot, and why a twelve-year, multi-billion-dollar effort should abruptly be abandoned on the rather ridiculous excuse that the public had become disinterested and non-supportive.

Indeed, the disinterested public was NEVER informed that we would not go back to the Moon. Instead, the Moon, fully supplied with water and atmosphere, was simply caused to fade away into official oblivion.

And there the matter would have rested except for the emergence of unofficial versions of the Moon, its anomalies, and its curiosities and all of which have turned out to be correct regarding lunar water and atmosphere.

If one takes the interest and time to read some of these unofficial sources (perhaps beginning with Fred Steckling's competent 1981 book), then one possible reason hooves into view.

As but two historical examples that help give reality to this reason, the following photos (acquired during the 1960s and identified here by NASA reference numbers) unambiguously show "airborne" objects near the lunar surface:

NASA Apollo 11 photo, No. 11-37-5438 clearly showing a luminous cylindrical-shaped object in flight above the lunar surface and exhibiting a high-altitude contrail.

NASA Apollo 16 photo, No. 16-19238 clearly showing a rather enormous, luminous cigar-shaped or cylindrical object casting its shadow on the lunar surface.

The cylindrical object in the NASA photo takes on added interest for the following reason. During September 1998, the cable station TNT aired a quite good documentary entitled "Secrets of KGB UFO Files." It contained some especially impressive footage (acquired circa the late 1960s) of Soviet MIGs encountering UFOs.

The footage was acquired via nose-cameras of the MIGs sent aloft to intercept unidentified objects intruding into Soviet air space. Among the several UFOs photographed by nose-cameras was a long cylindrical object moving rapidly above an Earthside cloud formation.

When the object sensed it was spotted, it rapidly put on speed and vastly outdistanced the MIGs chasing it. The TNT documentary indicated that the object had to reach speed of MACH-3 in order to do this. MACH-3 is very fast, and no Earthside craft is anywhere near capable of it.

The size of the fast-moving cylinder was estimated to be two or three times that of the Soviet MIG craft. It almost exactly matches the one in the earlier NASA photo, but which seems to have been much larger.

But we need not lean on historical NASA photos for evidence of this kind, or even on past unofficial versions or books.

Turn to the Internet, and especially to the weekly UFO ROUNDUP which provides a day-by-day listing of Earthside UFO sightings that are reported world-wide to this remarkable Internet publication.

Or access CNI News, a twice-monthly Internet news journal addressing UFO phenomena, space exploration and related issues.

These two excellent Internet sources reveal an almost obscenely large number of UFO cylinders, cigar-shapes, triangles, boomerangs, discs some of them luminous. All of them are quite busy doing whatever they are doing in Earthside's atmosphere, and sometimes just above treetops.

After the single, most obvious implication of these Internet sources sinks in, IF it does, one might wonder why luminous UFOs are found in the vicinity of both Earth and the Moon.

Another Internet approach is to access the general topics of MOON or MOON STRUCTURES or MOON BASES in the Internet's search engines. One can come across, for example, an article entitled "Astonishing Intelligence Artifacts Found On Mysterious Far Side of the Moon," authored by Jeff Rense, (with computer enhancements by Liz Edwards of Wonder Productions).

Indeed, the marvelous search engines of the Internet will lead one thither and yon through all kinds of lunar facts and factoids of which only one-tenth are needed to help fill out a very probable reason for NOT going back to the Moon.

That reason is more awesome than six billion tons of lunar water in the form of sub-surface ice.

Apparently that reason has been existing for a long time, was discovered to be existing during the lunar adventures and misadventures of the 1960s and is still existing today. And behind all the official scenes and cover-ups, that reason seems to be getting more complicated and extensive than ever before. And it is both ridiculous and hilarious that mainstream officialdom still pretends it doesn't exist, and still struggles to maintain the cover-up.

One of the strangest factors about all of this is that the topics of lunar UFOs and structures are seldom integrated into the overall Earthside UFO situation and its very many books and discussions. Indeed, in spite of copious lunar UFO evidence, Ufologists seems to avoid the Moon like the plague. As but one recent example, a new book came out in early 1998, entitled *UFO Headquarters: Investigations on Current Extraterrestrial Activity,* by Susan Wright.

This book is quite nice because for those not saturated with the UFO information available, it briefly reduces massive amounts of confusion into something easy to read and comprehend.

However, it makes no mention of the Moon even though there are very many available sources regarding it.

It would seem that the phrase EXTRATERRESTRIAL ACTIVITY might include ET lunar activity—in that if ETs do exist, then certainly getting to Earth's skies and to the Moon (and even perhaps colonizing it and its water) would not be impossible for them.

And IF they are loitering in the lunar environment, perhaps they can chase away well, NASA efforts, of all things.

SUBSCRIPT

MARS

Method of site acquisition:

Sealed envelope coupled with geographic coordinates.

The sealed envelope was given to the subject immediately prior to the interview. The envelope was not opened until after the interview. In the envelope was a 3 X 5 card with the following information:

The planet Mars.
Time of interest approximately 1 million years B.C.

Selected geographic coordinates, provided by the parties requesting the information, were verbally given to the subject during the interview.

From: Mars Exploration, May 22, 1984, CIA RDP96-00788R00190076001-9. Declassified on August 8, 2000.

INTRODUCTION

BY: STANLEY KRIPPNER, PHD[1]

I had the good fortune of meeting Ingo shortly after I moved to New York City in 1972. Even before coming to the "Big Apple," I had befriended Zelda Supple, the first full-frontal model to appear nude for Playboy magazine, and it was Zelda who introduced me to Ingo.

Zelda, Ingo, and I shared an interest in parapsychology, and Ingo claimed that he could project himself, or at least part of his psyche, to remote locations. Ingo said that he first recognized these abilities as a child. In 1971, he began to work with various parapsychologists in New York, including Dr. Gertrude Schmeidler with the City College of The City University of New York, and Dr. Karlis Osis and Dr. Janet Lee Mitchell with the American Society for Psychical Research (ASPR). With Dr. Schmeidler, Ingo appeared to demonstrate "psychokinesis," the alleged effects of his psyche on a distant object. With Dr. Osis and Dr. Mitchell while purportedly "out-of-the-body," Ingo correctly identified several small objects that had been placed in an enclosed space on the ceiling of the room in which he was located. While at the ASPR, Ingo coined the term "remote viewing" for this process.

During my visits to the Soviet Union, I had learned about high voltage photography, a process developed by Semyon and Valentina Kirlian to identify flaws in metal surfaces, such as the parts of airplanes. Other Americans were quick to claim that this procedure could capture the so-called "human aura," although the Soviet researchers told me that the electrical discharge would destroy any "subtle" bodily emanations. Nonetheless, any number of investigators published dramatic photos of human body parts surrounded by colorful emanations. "Blue Head," a colorful painting

---

[1] Dr. Krippner's biography and a bibliography of his work is provided in the Selected Bibliography section.

Ingo once gave me, shows the same bright radiations.

Several of Ingo's paintings seemed to show luminescence around the heads and bodies of human figures. Some of my students and associates noticed the similarity to "Kirlian photographs" and asked to obtain Ingo's permission to engage in a research study. Ingo graciously agreed, and our group photographed his right index fingertip with a low frequency photography device during several conditions. They included his ordinary waking state, his alleged "out-of-body" state, and a state during which he imagined heat emanating from his fingertip. Three photographs were taken in each condition.

Although no firm conclusions could be drawn from such a small sample and, admittedly, informal conditions, it was apparent that the three sets of coronas differed in remarkable ways. The emanations appeared to be denser when Ingo was "out-of-the-body." When he imagined heat, the photographs showed flares and breaks in the corona. From my perspective, these results were not surprising because body temperature, finger pressure, and perspiration are among the variables reflected in the coronas.

Ingo congratulated the research team for their efficiency and professionalism, compliments that were well received. Daniel Rubin, a member of the research team, and I published the photographs in our co-edited book, *Galaxies of Life: The Human Aura in Acupuncture and Kirlian Photography*, published in 1973.

Ingo wrote several books of his own, including *The Great Apparitions of Mary* (1996), a work remarkable in its scholarship, providing anthropological, historical, psychological, and theological perspectives. His chapter on Medjugorje, Yugoslavia, was of special interest to me as I had visited the village and had interviewed some of the young people who claimed to have encountered Mother Mary on a regular basis. Ingo autographed my copy of the book, stating, "We met way back and have lived through it all." He also inscribed his incredible novel, *Star Fire* (1978) to me as "a maker of dreams."

One of Ingo's most intriguing books was *Psychic Sexuality* (1999). He made no secret of sexual inclinations, but told me that he had been celibate for several years and intended to remain so for the rest of his life. On one of my visits to his studio, he showed me his collection of collages featuring nude male figures. Far from being pornographic, this artwork was painstakingly crafted, sometimes with five or six layers skillfully overlapping. Although done for his own enjoyment—and perhaps sublimation—these pieces ended up at the Leslie Lohman Museum in Manhattan, where they are highly regarded for their craftsmanship. His major metaphysical artwork is at The American Visionary Art Museum in Baltimore, Maryland, while his archives on consciousness studies are at the University of West Georgia,[2] which holds the collected papers of us both.

His work with Dr. Schmeidler and at the ASPR brought him an invitation to participate in experiments at the Stanford Research Institute in California, but he soon tired of simple card-guessing tests. He told the researchers that he could "see" anything on the planet, and they took him at his word. Soon, Ingo and the researchers Russell Targ and Harold Puthoff were talking about "remote viewing," a term that quickly entered the parapsychological lexicon.

Soon, Ingo would return from California, telling us about "what I did at Stanford." He would discuss his effects on a magnetometer and his accurate description of a site for which he had been given only the latitude and longitude coordinates. Once The Central Intelligence Agency took an interest in Ingo, there were no more anecdotes. However, he agreed to undergo neuropsychological tests at the Canadian laboratory of my colleague Michael Persinger who observed brain changes during Ingo's self-induced shifts in consciousness. The results were published in the prestigious *Journal of Neuropsychiatry and Clinical Neuroscience.*

I was delighted when Ingo invited my wife, Lelie, and me to be present when he and Harold Sherman (working from his home in Arkansas) engaged in the adventure to remote view Mars described in "9: A Psychic Probe of Mars." Lelie and I arrived early so as to be accustomed to the environment and the other observers. The event is so well described in this chapter that I cannot add anything of

[2] See the notes on the University of West Georgia in the Selected Bibliography section.

value. However, both Lelie and me were surprised at how casual Ingo was during the session. He was dressed informally and held a small cigar that he rarely placed to his lips. It only took him a few minutes to reach his purported destination, and then gave articulate descriptions of what he "viewed" so remotely.

Time will account for the accuracy of Ingo and Harold's reports on their remote viewing experiences of Mars. However, Ingo never claimed to have a perfect record. In recalling his work with police departments, he claimed success in only three of the 25 attempts to resolve a case. Ingo's personal style was modest yet confident, profound but humorous. I enjoyed one last visit to his Bowery studio before his death at the age of 79 when, to cite the title of one of his books, **he kissed Earth goodbye**.

BY: JANET LEE MITCHELL, PHD[3]

I first met Ingo Swann when he came to an open house in 1971 at the American Society for Psychical Research in New York City where I worked as a research assistant. He was extremely kind, an intelligent and interesting gentleman. We agreed that we were creative invisible beings operating through bodies. At a time when women weren't taken very seriously in a professional way, he treated me as an equal partner on projects and I considered us co-experimenters.

We first worked with perceptual tests on what I called out-of-body vision and he called exterior vision, but we didn't let terminology interfere with discovery. Our mutual purpose was to find out how extrasensory abilities work and share that information with others. I worked with Ingo on many different projects for about 20 years.

I recorded his perceptions in the laboratory, in the field as he looked for resources underground, and in 1974, he ventured out to the planet Mercury to see what he could before Mariner 10 got there and radioed back data on the relatively unknown planet. This was the early learning stage of his abilities and there was considerable success in all efforts.

In 1975 and 1976, I recorded two sessions where Ingo (in New York City) and Harold Sherman (in Arkansas) projected their vision to the planet Mars. These records were notarized and sent to various laboratories. They are now in the archives at the University of West Georgia[4] along with actual tape recordings of the responses. You can see for yourself how these coincide with what is reported by Ingo in his "9: The Psychic Probe of Mars."

---

[3] Dr. Mitchell's biography and a bibliography of her work is provided in the Selected Bibliography section.
[4] More information is provided in the notes on the University of West Georgia in the Selected Bibliography section.

"9"

THE PSYCHIC PROBE OF MARS

BY: INGO SWANN[5]

During 1975, NASA's Viking orbiter spacecraft was on its way to the planet Mars, destined to arrive there in 1976 and pull off a touchdown on its surface. Mars is the infamous "red" planet, the legendary home of the god of War. It is next from Earth outward from the Sun. In between Earth and Mars is the asteroid "belt" which many believe is composed of the remains of a shattered planet now circling the Sun in a gigantic ring composed of dust, ice, rocks, and asteroids, some of which are quite large.

Although strange phenomena had variously been observed on Mars since the invention of the telescope, prevailing scientific opinion held that it was a "dead" planet entirely inhospitable to life. Mars is just over half the diameter of Earth, but has only 11 percent of Earth's mass. It has a thin atmosphere composed mainly of carbon dioxide, in which spectroscope analysis had detected small amounts of water vapor, ammonia, and methane. It is also very cold on Mars, its temperature varying from about 80° Fahrenheit at noon to about -100° Fahrenheit at midnight.

In 1877, the Italian scientist, G. V. Schiaparelli, noted the existence on the Martian surface of strange markings which he referred to as canal 1 ("channels" or "canals"), and the later astronomer, Sir Percival Lowell (1855-1916), created a lasting controversy by accepting these "canals" as the work of intelligent beings. Subsequent scientific analysis from Earth of Mars, however,

---

[5] Found with Ingo's draft materials for *Penetration* at the University of West Georgia, written most likely in 1998.

discredited Lowell in establishing that the strange markings were some sort of natural feature, perhaps similar to rock faults on Earth. And there the matter lay until 1975 and 1979 respectively.

In 1975, with some modicum of interplanetary remote viewing success under our belts, Harold Sherman and I planned a Mars probe in advance of the first Viking touchdown. There were certain difficulties with regard to Mars which differed considerable from those of Jupiter and Mercury. A great deal <u>was</u> known about Martian conditions, and lest Harold and I merely repeat what was already known, we decided to read up on the planet and become "grounded" in existing knowledge about it.

In research, this reading up about a target in advance of psychically "travelling" to it is called "front-loading." When, in the practical application of remote viewing, clients wanted a distant site "looked at" they were obviously interested in what was <u>not</u> known or hard to find out about it. Most of the usual intelligence-gathering methods are very good, but their efficiency is dependent on "penetrating" into areas of which the usual methods have little or no access. The remote viewer front-loads because he or she (1) does not want to waste time providing information that is already known, and (2) the front-loading will make it easier for the remote viewer to recognize what is strange or unexpected about the site. What Harold and I would be "looking" for in the Martin environments were phenomena not presently know by science, and to determine this we had to front-load ourselves with what was known about the planet.

Our Martian studies took about three months, but by the first week in June 1975 both of us felt ready and geared up to penetrate Mars and its environments. At the time I was on vacation from the tedious work at SRI, and so with Harold at his home in Arkansas and myself in New York, we agreed to go to Mars at 9:00 p.m. EST on 14 June 1975. The reason for this late hour was again determined by the "directions," which we derived from the astrological ephemeris. At that late hour, Mars would be just ahead of the Sun, which had just set in the West. Relative to Earth, the Sun was "in" the zodiac sign of Cancer, with Mars "in" Taurus." This effectively placed Mars on the other side of the Sun, which is to say a very distant from

Earth. In any event," I told Harold, "just 'go' to the other side of the Sun and then some. If we lose our way out there, just 'go' to the tiny red planet. There's only one red planet in our solar system, and so it will be Mars."

Since I was in New York, and not in California with the rest of my colleagues, I invited Dr. Janet Mitchell to observe as formal recorder of the information, and as additional observers Dr. and Mrs. Stanley Krippner, both well-known in parapsychological research, and Mrs. Zelda Supplee, a close friend.

It took about eight minutes to sight Mars psychically, and Harold later wondered, too, why it had taken him so long, since most psychic impressions are almost instantaneous. During this eight-minute period, for the first time I had the explicitly distinct experience of "travelling," and in fact the room in which I was sitting "disappeared" completely along with my body—which defeated any sketching I had planned to do. I could hear myself talking, though, which was being recorded. The typed notes of this talking ultimately extended through five single-spaced pages. But of all the data accumulated, only one set of them attracted much interest at the time. When the experiment was over I was flabbergasted. But at that moment, the telephone rang. It was Harold Sherman calling from Arkansas.

Now, Harold Sherman was an easy-going, gentle person, with interior strengths to be sure. But I had never heard him shout. And this he now did. "Did you see what I saw!" he shouted. I was still somewhat in awe of what I had seen that I responded vaguely: "Well, if you mean what no one will believe, then yes!"

There was a long moment of telephonic silence between Arkansas and New York. "If we release this, Ingo," Harold warned sternly, "people will be sure we are psychic kooks. Perhaps we should edit that part out of our impressions."

I considered. Then: "Harold, I've got Krippner and Mitchell here. They observed and heard my end of it. If we edit, then all our other probes will be suspect. So we cannot, and must sink or swim with what we saw."

"Oh Boy!" Harold groaned. "But you're right, of course."

I was thinking fast, though. "Harold, what about a compromise.

Let's not openly circulate this Mars stuff. We'll both get our tapes typed up and have every page of them notarized for posterity, and in this way we will have a complete notarized statement." And so all parties agreed.

To get to the nut here, Harold had seen: "As I descend [to the Martian surface] for a closer view, I see—can it be possible—domelike structures, great mounds which give evidence of some form of intelligent life."

And: "... I get the feeling a higher form of life has existed [here on Mars] or does exist and may have gone underground. And now I see some objects in the sky some miles above the surface. They [the objects] seem to be either moons or satellites of some sort. I see at least five. They appear to be stationary."

I had seen: "I have come upon something which I guess I will really have to go out on a limb to talk about—because it looks like one of those towers that carry high voltage lines. It is man-made and it is structured and it is bent and no longer in use, but man-made! I wonder if that might not have been some kind of homing device like airplanes use these days when they go from one continent to the next. It is built out of something that doesn't rust, but it is bent."

And: "I was tempted to say [that] back in that crater that I [just] passed over, something that looks like squares, maybe 100 feet [large], like a checkerboard. That's a rough thing to say, but those look like they're man-made. But ruins, old."

Dr. Stanley Krippner recovered his composure and on an extension telephone conducted an after-session "debriefing" of Harold and myself in which it was established that the "squares" or "domes" seemed like what would be seen of "old Roman ruins," but that the bent structure and the satellites were "newer."

After this, Harold said he had to "go to bed to recover and contemplate the meaning of all this," and those of us gathered in New York opened a couple bottles of wine and sat silently. Very silently.

In this way, both Harold and I finally encountered our cowardice level. After all, "man-made" structures on Mars were "impossible," and we were really verging on joining the "irrational set" that all anti-psychic skeptics liked to spend considerable time debunking.

Thus, the "Mars psychic data," although duly notarized on every page, was "retired." The gossip line, however, had not retired, and soon Harold and I were receiving telephone calls asking us to describe the "buildings on Mars." We had both agreed to offer "No Comment" statements. And that seemed to be that, and we awaited to learn whether or not the Viking lander would discover buildings on Mars. Which in fact it did not, since NASA had decided to land Viking in a small crater rather than on a high plateau where it might have sent back pictures of a wide variety of landscapes. All that was visible in the crater were a variety rocks and rock crater walls which cost several million dollars to photograph.

But all was not lost with regard to feedback of this extraordinary psychic probe.[6]

***

[6] Psychic Probe of Mars 1, June 14, 1975, Responses of Harold Sherman, and Ingo Swann, as found in The Archives of Harold Sherman, Torreyson Library, Archives and Special Collections, University of Central Arkansas, have been reproduced in their original form on the following pages. On January 29, 1976, Harold Sherman and Ingo Swann undertook a second remote viewing session of Mars. Present were Edwin (Ed) May and Dr. Janet Mitchell. This second session was not discussed in Ingo's "9," but is referenced in Dr. Mitchell's Introduction. These documents, as found in The Archives of Harold Sherman, Torreyson Library, Archives and Special Collections, University of Central Arkansas, have been reproduced in their original form following the 1975 session documents. See the notes on the University of Central Arkansas in the Selected Bibliography section.

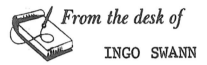

From the desk of

INGO SWANN

Dear harold — This
is TOP SECRET —

Russell Targ came
through with a NASA
official who is going to
help us on this. We
don't want to spoil this
by premature announceme-
nts. Love Ingo

~~Dear Harold~~

PSYCHIC PROBE OF MARS I

June 14, 1975

Harold Sherman Response

10:10 PM EST

As I withdraw my conscious mind's attention from my physical body and its immediate surroundings and enter the field of my inner consciousness, I seem to take on an expanded awareness. It is my conviction that an element of mind makes instant contact with whatever point or objective is visualized and that if the mind's attention can be maintained at that point, one can mentally see or sense to a degree what is taking place there. This time the assignment is for me to attempt to receive impressions from the planet Mars.

As I so direct my mind, I am not conscious of any travelling sensation between Earth and Mars. I just suddenly seem to be in the vicinity of the planet surrounded by a reddish glow or atmospheric haze. There are enormous whitish cloud banks fantastically pitted by this glow as I find myself feeling like a particle in space descending through the cloud cover. The clouds are moving rapidly, leaving great spaces between them, which permits glimpses of the surface of Mars far below. This is a spectacular sight, some volcanoes miles high with red flames erupting and atomic-like clouds boiling up above. I sense currents, magnetic currents or energies swirling about me and I am conscious of the interplay of varied-colored lightning flashes in and through the clouds. I am still high above the planet and as I come beneath the cloud cover, I see that great stretches of surface are cratered. In these areas, there are mammoth dust storms.

For the first time I get a sensing of Ingo Swann beside me. We seem to be looking at a gigantic strip of green vegetation hundreds of miles wide and possibly several thousand miles long. A valley of immense size and extent around the equatorial belly of the planet. There are clouds of water vapor rising over the fields of vegetation. This area seems to be bounded by chain-like links of rugged black mountains.

There are definitely some forms of life on Mars. Over the body of the planet I sense a thin atmosphere, some moisture and a warm temperature which could support life. But I also

WITNESSED SIGNATURE June 16, 1975

Edgar Lee Mitchell

BERTRAM E. STRAUSS
Notary Public, State of New York
No. 24-216120
Qualified in Kings County
Commission Expires March 30, 1977

sense frigid atmosphere in the north and south polar regions.
I am most impressed by a panoramic view of Mars and the lush
green land extending in a wide band as far as the mind's eye
can see, with whitish capped mountains in the far distance.
The land areas on the planet, as a whole, look like they
contain great plateaus with the mountain peaks and volcanic
cones rising out of them. The plateaus are almost as though
they have been geometrically laid out.

As I descend still farther for a closer view, I see --
can it be possible -- domelike structures, great mounds which
give evidence of some form of intelligent life. There must be
some water, but I don't see bodies of water as such. I got the
feeling that life has gone largely underground or is sheltered
in some way. But I get the feeling a higher form of life has
existed or does exist and now I see some objects in the sky
some miles above the surface. They seem to be either moons
or satellites of some sort. I see at least five. They appear
to be stationary. Can it be that the atmosphere is rare and
that life has had to adapt to a somewhat different environment
in these areas, as Mars had more water at one time? And, is
this one of the problems of any life now existing on the planet?
I have no knowledge of chemistry or physics, but there seems to
be a scarcity of some elements or lack of balance even though
conditions are still such as to permit some forms of life.
That's it.

WITNESSED SIGNATURE January 18, 1971.

[signature]
MIRIAM E. STARKER
Notary Public, State of New York
No. 24-9216120
Qualified in Kings County
Commission Expires March 30, 1971

[signature: Janet Lee Mitchell]

PSYCHIC PROBE OF MARS I   June 14, 1975

Discussion between Dr. Stanley Krippner (K), Harold Sherman (S)
and Ingo Swann (I) after all responses had been recorded.

K: There were several correspondences between what you (S)
    just said and what Ingo told me so when you get the full
    report you will be very delighted to see the points of
    similarity that you both made.

S: Yes. I made some extra recording here. I can be wrong,
    but I recorded, "I seem to see Ingo stretched out on a
    davenport with Stanley and Janet facing him in the front
    room of the studio. He appears to be talking into a tape
    recorder as the impressions come. Janet, with a note pad,
    is to the left and Stanley to the right from Ingo's position
    on the davenport."

K: That is exactly correct except Ingo was in a chair, not a
    davenport. But you're right. From where Ingo was sitting,
    Janet would have been on the left and I would have been on
    the right and we were both facing him.

S: That is exactly my impression.

K: And, we were in the front room and he was speaking into a
    tape recorder. The tape recorder was directly in front of
    him and Janet was taking notes and I was not.

S: That's very interesting. Then I said, "Ingo's eyes are
    closed for the most part."

K: Right.

S: "He sighs deeply occasionally."

K: Yes.

S: "Starts to speak at times, stops to correct himself or
    changes comments as different impressions come. As I
    return from my fixation on Mars, I seem to stop off at
    Ingo's as though magnetically attracted because of our
    dual concentration on this project." Did the phone ring
    during Ingo's session or did someone take it off the hook?

MARS I - Post-session discussion   2

K: It was off the hook. But there have been noises outside. There is a fire siren right now and any number of other things.

S: Stanley, apparently my impressions are almost exact with respect to your situation there.

K: Oh yes, it certainly was. Would you like to say hello to Ingo?

S: Yes.

I: Hi.

S: Hello Ingo, how do you think it worked out?

I: I didn't hear yours yet.

K: Yes, let me just mention a couple of things. Both of you were involved in dust storms, craters, vegetation, slight atmosphere, winds, man-made objects.

I: What did you see?

S: Ingo is hearing this for the first time?

I: Yes, what did you see, Harold?

S: I saw some objects in the sky like moons and satellites and I saw some domelike structures on the ground.

I: You did? OK.

S: I can hardly wait to hear Ingo's impressions.

K: I think there are two discrepancies that I picked up. By that I mean things you reported, Harold, that Ingo did not report. Harold, you reported these objects in the sky and Ingo made a point of saying that he couldn't even see the two moons which we know exist around Mars. But maybe you were on different sides of the planet.

[signatures]

witnessed Signature
June 6, 1973

HARTRAM E. STRAUBEL
Notary Public, State of New York
No. 24-8216120
Qualified in Kings County
Commission Expires March 30, 1974

MARS I - Post-session discussion   3

S: This is definitely going to be interesting for you people
   to evaluate. Did I see vegetation to a degree like Ingo saw?

K: Oh yes, he saw a lot of vegetation, but he did not see the
   erupting volcanoes that you saw. Other than that, I think
   there was quite a bit of congruence.

S: This was a good check on veracity, I believe, when I seemed
   to come back and found myself mentally in the studio with
   you. If my impressions were exact then, then I kind of
   like to check every once in awhile.

K: Sure, right.

S: Thank heaven you were there to verify these impressions
   and they were recorded too, weren't they?

K: Yes.

S: We're going to have to work on Venus next.

K: Right.

I: Harold, we'll go back to Mars in about two or three months
   when we get all our data lined up. Dr. Puthoff is going to
   send us the exact things that the Viking Mission is going to
   look for and Janet already got a good write-up which I
   haven't read yet. I didn't want to do any more so we'll
   get all that together and maybe we can go again when we
   have these things down and see what's what.

S: Yes. Well, I had the impression I was right there with you
   people.

I: Yes. I was with you when you **took off, I** think.

S: I felt so.

I: Were you out in your little house out there?

S: Yes, I was in the little annex.

MARS I - Post-session discussion   4

I: And you were lying on the couch?

S: No, I was sitting in a chair.

I: OK. Well, I knew you were out there anyway.

S: I laid on the couch for just a minute after I finished. I
    finished about ten minutes before nine (CST).

I: I see. So did I. I was very tired.

S: I was, too.

I: OK. We;ll get it typed up and sent out to you fast.

Those present in Ingo's studio during the probe were:

>   Dr. & Mrs. Stanley Krippner
>   Ms. Zelda Suplee
>   Mr. Ingo Swann
>   Ms. Janet Mitchell

Janet Lee Mitchell

WITNESSED SIGNATURE
June 15,1975

WILLIAM E. STRAUCH
Notary Public, State of New York
No. 24-2161130
Qualified in Kings County
Commission Expires March 30, 1976

PSYCHIC PROBE OF MARS I   June 14, 1975

Ingo Swann Response   9 - 9:55 PM EST

        When you try to leave Earth, I guess the belts around
sort of take hold. Anyhow, I have risen out of the night side
of Earth and I see the Sun and I know I have to go on the other
side of the Sun to find Mars because Earth and Mars are at their
farthest, almost at their farthest proximity from each other. I
don't know why they call the Sun a yellow star because it looks
pretty white. It has an enormous halo which looks lavender.

        Mars looks like a fusia colored pinpoint. I thought it
would look orange. It's along way away. I seem to sense extending
from that, I guess it's Mars, an enormous, what would you say, a
magnetic penumbra. Is there such a thing as negative magnetism?
Anyhow sort of extending enormously far out from the planet is
sort of a magnetic condition which must polarize the sunlight
just a little bit. So when you get inside that, the Sun looks
a little darker. There we go. This magnetic shell, call it a
shell, all around the planet must be at least about 350,000
miles outside the planet. The magnetosphere around Mercury
was very active and the contrast to Mars, which seems to be
very peaceful. I guess I'm approaching Mars on its day side
and it's now turning -- it's losing the deep color it had and
it's now turning pink or maybe a golden yellow.

        Compared to Earth, it has to be composed -- it seems
more dense anyway. It's strange, I don't see either moon.
They must be on the other side somewhere. That must be the
upper atmosphere which is extremely cold.

        I get the impression of -- these must be dust particles
which, as I go closer now to the surface, obscure it and make
it hazy. I'm not sure what a convection current would be, but
it appears that there are air currents maybe like ocean currents
here on Earth but these are in the air. In other words, you
have one area where there is dust in the atmosphere but it is
not moving too fast and in other places it is moving like a
long current.

        There's an intense softness like if you made a sound
on Mars, it might get stopped like a soundproof room here.

MARS I - Swann Response   . 2

The, I guess, audio frequencies may get absorbed by the heavy
dust in the atmosphere. The area where I got to is filled with
dust. Let's see now, how do I move to somewhere else? Some of
these currents move almost in straight lines it looks like. Why
should that be? It's almost as if in the air, atmosphere, there
are extreme high and low pressure areas which tend to discharge
toward each other creating those currents that can almost go on
a straight line.

I just found what looks like a cliff, which contrasts
enormously to the dust. It is extremely hard of some sort,
like granite, but it isn't. Maybe that's like a volcanic con-
glomerate, something like that. Anyhow it seems yellowish with
black in it and it glitters here and there like it had silicates
in it. That was the cliff I bumped into. If I go up to the top
of the cliff there is a huge overdraft. I wonder what kind of
dust or sand it would be that when it's blowing around in the
air, it is very soft and fine and when it settles, it packs in
very hard right away? Anyhow, there is a lot of this. On top
of the cliff seems to be now a great plain and the wind blows
off the cliff and seems to cut long straight lines in this
heavier stuff. I guess this is an extreme wind erosion taking
place all the time and when it breaks the particles loose, they
suspend in the air very easily and when they do settle, they
pack in very hard. I follow the plain that looks very flat.

This area where I am is very dry. The surface in the
sunshine seems to be about 80°, but if you go up five feet,
it drops about 50° in the temperature already. I see in the
distance -- I have the impression I am now moving northwest.
In the distance I see some mountains which seem to be black or
reddish-brown, I guess. I guess a warm surface contrasted to --
if you go up 100 feet in the air, it is very cold. This might
give rise to these extreme, I want to call them convection
currents. I'm not sure what those are.

In these mountains I see that these must have been laid
down at one time as on top I can see various stratas and toward
the bottom I see a great big layer of what looks like sandstone
and it's carved out. It has scoops in it. That's from wind.

WITNESSED Signature   June 18, 1975

MARS I - Swann Response   3

Wind caverns. And then on top of that is of obviously a harder
substance which sticks out like a ledge over the other area.
Higher up on these mountains, once you get out of the dust, you
can hear the wind screaming, one of my favorite sounds. In
places those mountains look damp like the rocks are extruding
something. It doesn't seem like it is water; it seems like
something else. These mountains must be about 8,000 feet high,
very impressive. I'm going over them and I see a crater. First
one is very large. There are whirlwinds in the crater, but
hardly like the tornadoes we saw on Jupiter. I see maybe three
of them.

On the northwestern slope of the crater seems to be all
drifted in with sand dunes so I pass along that now and move
still northwest and there's another canyon, I guess, extremely
large, like ten miles across. Not a canyon, a sort of -- like
part of the whole surface has subsided a little bit. These
must be earthquake areas. A faultline, that's what it is.
Seismic area.

Then I see another range of mountains which are higher
than the last ones and look more pointed and I have the distinct
impression that in the shadowy side there are icicles. The last
thing I expected there. On the sunny side of these -- heats it
very warm. Where the sun doesn't heat, it looks like there is
a covering of ice and icicles here and there. I got on it from
the more gentle slope and (on the left) is lots of cliffs. Then
it's another plain. It looks like there has been water here
because the surface of the sand is dry and cracked like it was
once wet and now it is cracking.

I wonder if that dust that I first looked at might be
composed of not earth it all but spores? I've come to an area
now that looks like it has something like lichens but they are
reddish and maybe with green casts. This must be some sort of
a vegetable. A thing that grows one right after another like
on chains. Chains, there we go. Like nitella plants that I've
worked with, you know, just one after another and whey they
decompose, they break apart and go up in the air.

Witnessed Signature   June 18, 1973

WILLIAM B. STATE
Notary Public, State of New York
No. 24-9216120
Qualified in Kings County
Commission Expires March 30, 1976

David Lee Mitchell

It seems very trite to say that if I go now what must be farther north -- I hope I'm not in the south pole. Maybe I'm doing it backwards. How do you tell if it's north or south? It must be the north.

These lichens turn greenish farther on where there is, looks like it must be water crystals in the -- where the planet curves so that the Sun doesn't heat it so much, there must be water crystals and they form on water crystals.

I was tempted to say back in that crater that I passed over something that looks like squares, maybe 100 feet, like a checkerboard. That's a rough thing to say but those look like they're man-made, but ruins, old.

Note: The following does not appear on the tape but was recorded by Janet Mitchell in shorthand during the session. No information was lost.

(I was moving slower before but now I'm going to move faster. I just found a big canyon which goes down quite deep and the wind blows like hell through it. There is ice at the bottom in places. There are lots of little caves where the wind has dug them out. This might be the big rift valley that appears in our briefing data here. It is like a huge network of canyons and tributaries and the pressure seems to change in the bottom. Somehow I feel if you walked up and took a hammer and hit some of these cliffs they would reverberate, maybe like a bell. I'm almost getting a sensation or smell of petroleum. Mars is built like this (gestures outline of sharp cliffs and ledges) great dropdowns and wind, oh. There is something extremely strange about the geology. I can't quite even get it in focus, whatever it is. I seem to have gotten stuck down in the bottom of this valley. Up and out on the plain the wind is blowing.)

I have come upon something which I guess I will really have to go out on a limb to talk about it because it looks like one of these towers that carry high voltage lines. It is man-made and it is structured and it is bent and no longer in use, but man-made. I wonder if that might not have been some kind of a homing device like airplanes use these days when they go from one continent to the next or from one mountain range to the next. It is built out of something that doesn't rust, but it is bent.

Witnessed Signature June 4 1975

Janet Lee Mitchell

MARS I - Swann Response   5

I must have passed now around the planet and I'm coming
down into the southern hemisphere because I could see that big
plateau that is referred to in photographs with its huge volcanoes.
That's the first volcanoes I have seen. I see about twenty of them
now, I guess. Quite huge.

In this area I seem to see something that looks like Chinese
jade plants. They're little plants that are growing and look like
they have sacs. Do they contain water? I believe they do. I can
draw a picture of that later on.

I find it hard to understand why Mars looks reddish because
it doesn't look reddish; it looks yellowish and in the places where
the geology is showing through it looks blackish and dark brownish.
The general tone of the planet color-wise is sort of a buff color,
but those other plants or whatever they were looked red and greenish
and a green tinge to them. These down here look yellowish. There's
some of the other ones, too. These must be little plants holding
little sacs of water maybe.

I seem to want to turn my attention to look at the atmosphere again. In this area there seems to be something that goes
on in the air. Lightning, them it is. But there are no clouds,
just lightning and it doesn't touch the ground. It might be
because there are enormous up-drafts along these volcanic ridges
here and it discharges.

I'll make a summary: It's a very dusty place, I sense
ice and the differences in pressure areas as they sort of discharge each other they create valleys through what looks like
a hard impacted stuff, but when it blows up in the air it turns
to very fine dust. I would say there must be spores from these
red kind of plants and the other kind of plant I saw is a little
sac thing that seems to store water. I saw two man-made structures
(which we better not print) and the noise is terrific. There is
lots of wind and bell-like sounds as the wind blows in and out of
these caves. It sounds like organ pipes music. Different sounds
in various places. The Sun looks smaller and there is this
strange kind of -- I'll call it a polarized something to the
light, which steps it down just a little bit. Maybe that is
why the planet looks red because if the light was reflecting
back from the planet in a polarized way which might just let
through the longer wavelengths. I think I'd like to stop now
and if Stanley has any questions, he can start.

PSYCHIC PROBE OF MARS I   June 14, 1975

Discussion between Dr. Stanley Krippner (K) and Ingo Swann (S)
after Swann response had been recorded.

K: I would like to ask you in what way these sensory impressions
came to you, such as the sounds and the sights which you just
described.

S: You sense a little enturbulation which doesn't go with the
rest of what one is looking at and so I turn my attention
to it and it gradually resolves into a sonic thing or a
light radiation thing or the surface must -- for instance,
absorb the long rays from the Sun but it must push them back
up too and that is what creates this intense -- this warm
thing near the surface that you go up even ten feet and it
is dropped 50° already. You know, you become aware of those
differences by sensing them, I guess. I know I didn't answer
your qestion.

K: A related question involves the sense you have of distance
from these features like you were in a valley one time, you
were in a crater once. Now did you actually sense closeness
to these geological features or were you at a distance sort
of vicariously thinking of what it would be like to be in
these geographical locations?

S: No, in just moving across the planet's surface psychically,
you see something that interests you so you go to it to
contact where the geology is exposed in these cliffs -- to
try to touch it psychically and you can say that there is a
soft, maybe sedementary, layer down here which the wind
scoops out underneath and this looks like a volcanic type of
stuff up here and you sort of touch it psychically to get a
feel of its composition. Those things break off, by the way.
Whenever you touch them, they break off and they have land-
slides.

K: Also, what was you time sense during this? Did you actually
feel the time was ongoing when you were on Mars?

S: The time seemed linear when I was moving, but when I was stop-
ping to look at something, it seemed sort of timeless.

K: In what way do you sense the color when you are psychically
travelling over Mars?

MARS I - Krippner/Swann Discussion   2

S: Well, I went there expecting it to be red and it wasn't
   red; it was some other color. Like it looked fusia from
   a distance. We are all taught to see that it is an orange
   or red planet and it is not. It is just a funny thing that
   goes on with the light as it bounces back out and so I was
   aware that it wasn't red and I had to decide on what color
   it would be and this appears to be sort of a buff color
   with intense contrast where the geology pokes up and where
   it is cut down.

K: Now, as you came into contact with these two objects that
   you felt were made by humans; did they seem very, very old;
   did they seem contemporary?

S: The squares -- if you were to go to an old Roman site where
   you would see just what was left or something, that is how
   it would be. But the tower -- I don't know how old it would
   be. It doesn't look weather-worn at all, but it is bent.
   It isn't used anymore.

K: Then the plant life that you described -- is this something
   that only occurs occasionally on the planet or does it exist
   in profusion?

S: It seems that must be a whole cycle where the plant life --
   I was wondering if it's maybe those red things anyway. The
   first things I described may be a form of animal/plant life.
   I remember from my studies in biology, we have something
   called mycetozoa in biology and the same plants are called
   myxomycetes in zoology. They are half plant and half animal.
   They go on. These look like they reproduce like this and
   then when they die, they all break up into spores, all the
   little cells, and these float through the atmosphere and
   when there's, I guess, a condition of -- they can't live
   where the Sun is heating the soil so they live along the
   fringes where it is cool enough for them to do it and
   that's where they grow. They seem to be associated with
   the polar caps or what scientists have observed as polar
   caps, but I don't think that they particularly feed on those
   water vapors there. They probably just thrive where it is
   not as hot. When it gets hot, then they dry out and all
   these spores go on and they're all over the planet it looks

S: like to me and they just take hold in this one special
   place where the Sun isn't heating the soil too much. In
   other words, it is just the reverse of heating on the
   water from the polar caps but its a retreat from the Sun,
   from the heat.

K: OK. Did you see anything that looked like canals to you?

S: No. Only where these -- sometimes these convection currents
   seem to go straight through the dust-laden atmosphere and
   create sort of a shift and look like straight lines. The
   other thing is where the wind has been blowing along this
   plain and it has cut dikes through that are quite straight.
   I didn't see any canals.

K: OK. Thank you. That's all the questions I have.

TARGS I - JUNE 14, 1775
SWANN DRAWING

life one

This seems to be
a plant/animal colony,
breaks up into dust-
like spores.

color red —
like blue red —
with green edges,
if they are healthy
and not drying
and germinating.
Turn brown when
the sporate.

life two

water sacs
plant.

Sort of whitish
opalescent, or
pearl green.

alternated cactus

PSYCHIC PROBE OF MARS II   January 29, 1976

Harold Sherman Response   11:05 PM EST

    In undertaking this psychic probe of Mars in association
with Ingo Swann, I am programming my mind in an attempt to zero
in on the planned landing sites of Viking A and Viking B, both
now on route to Mars. It is my understanding that Viking A is
scheduled to make a soft landing on Mars this coming July 4, 1976
and that two sites have been selected for each Viking in the event
that one or the other is not available due to storm or malfunction.
Since I have found that the moment I attach my mind's attention on
a specific objective, I seem to make contact at that point, I intend
to concentrate on each location independently and record any impressions of what I seem to see and feel there and after a minute's
relaxation, proceed to the next site until all four Viking landing
areas have been covered.

    Viking A, landing site no. 1 (19.5° N, 34° W): I feel as
though I am with the lander as it has just touched down. I am
surrounded by a heavy swirl of dust and as it clears, I find myself in a seemingly deep ravine. It is the floor of an immense
valley. I can see ridge-like rock formations aboùt me, dark in
color. The soil appears to be reddish-brown. It is hard and dry.
The atmosphere is thin but even so, denser than might have been
expected. Temperature is hot. As I scan the area I can see quite
some distance in every direction, save one, which appears to be
flocked by a rugged slope some quarter mile or so away. I seem
to be following the swing of the camera and the blocked area is
behind me to my left. The valley on other sides of me has spots
of vegetation. It has a harder look like a desert cactus plant.
I don't see any trees. The landscape has dry rivulets and deep
channel-like depressions which extend into the horizon rimmed by
high elevations or cliffs.

    Viking A, landing site no. 2 (20° N, 252° W): I am down
on what appears to be a large plateau with crater-like depressions
in the distance. The ground has large lava-like formations stretching
as far as my inner sight can see. The soil here appears dark and
semi-solid as though it could be shàpped and broken off. It is
dust covered in areas where the eroding wind as not blown it away.
There may have been water over this region many years ago. I see
no water now, in fact, I feel there is a scarcity of water on the
planet at this time. Despite this, there is a semblance of vegetable
growth, patches of green and yellow and there appear to be seasons
when vegetation is brought to life by water vapor due to temperature
changes and chemicals in the atmosphere. I have no knowledge of
biology or physics, so my observations will have to be translated
accordingly.

Mars II - Sherman Response   2

     Viking B, landing site no. 1 (44.3° N, 10° W): It seems
that this lander is to touch down in an extensive crater bed,
not too far from a number of volcanic cones that project upward
in viewing the landscape as the lander would be jettisoning down.
It looks like a gigantic pin cushion. This area is more spectacular
than former viewings. There have been lava flows in the distant
past, which have left large ribbon-like marks on Mars' surface.
Again, there is some evidence of plant life and green and reddish
clumps of vegetation. I see no evidence of animal life, although
I have a feeling that small forms could exist in some areas.
Visibility, when not obscured by strong gusts of wind which
sweep the region, extends for several miles, perhaps ten. There
are mountains in the distance and one volcano, which seems to be
active since there is a reddish glow and atomic-shaped clouds
above it.

     Viking B, landing site no. 2 (44.2° N, 110° W): Not too
unlike the appearance of the first site except it seems to be
closer to the volcanic area where there is every evidence of the
occurrence of violent activity in the long distant past. Craters
of different sizes spotted throughout in checkered landscape. The
atmosphere seems to be heavier and I have the impression that it
is laden with carbon dioxide. Some vegetation, again, to be seen.
Oxygen layers not too strong. I seem to feel occasional earth
tremors.

     A general survey of the planet Mars: I now direct my
mind to a free viewing of the planet Mars. My objective is to
look for possible signs of animal life. I have already noted
evidence of plant life. I am descending through miles of turbulent
cloud banks penetrated by vivid lightening flashes. As I approach
the surface of Mars, I am conscious of the presence of powerful
magnetic energies which give me the feeling that Mars is a laboratory
of mixing gases and chemicals. Untold quantities of charged particles.
I am emerging from the cloud layers over what appears to be the
north polar region, entering an area of intense cold, looking down
on a dust-covered and ice-encrusted region tortured by winds of
high velocity. I am strangely not buffeted by these winds, which
seem to scream past me. The white-capped stretches of land are
domelike in places. I have the feeling that a large and immense
supply of water has been trapped in the ice and is given off at
times in seasonal water vapor. The water has found its way into
the interior of the planet. It is pretty much solidified and is

seldom to be detected on the surface. It is contributing none-
the less to such life as can exist on Mars. There appears to be
a large reservoir of frozen carbon dioxide in both north and south
polar regions with perhaps a bit more in the north. This quantity
of carbon dioxide varies with the planetary seasons. An abundance
of frozen water exists in the polar caps, as well as gases.

I have the feeling as ancient environmental changes took
place on Mars that humanoid types of intelligent life upon it
tried to melt and drain off and channel this water in canals or
old riverbeds for use in growing foodstuffs for survival. The
Mars inhabitants at that time were forced to go underground to
escape the severe temperature changes and the surface drought.
I feel that the former existence of a highly intelligent civilization may be discovered when manned flights eventually reach Mars.
I am shown images of Mars dwellers who appear to be short, rather
heavyset creatures from four to five feet in height, possessing
large lung areas, large nostrils and a larger head, with thick,
rather rough, leathery skin, dark red in color. The body structure
seems to have been required by nature to suit the environment. I
would be surprised if any of these so-called Martians have survived
to this day. If they have gone underground with their scientific
knowledge and equipment and some of them have learned to cope with
the changing conditions on Mars, they might be expected to endeavor
to colonize on some other planet. This is mere speculation on my
part and not an impression in this case.

There is a strong deposit of iron on Mars which can be
related to the core of Mars itself. It seems to be surrounded,
even permeated by a magnetic field which unites with a magnetic
field of the sun and is associated with the force of gravitation.
Originally, I am impressed that the vortéces brought about by
these magnetic energies attracted the gaseous particles which
formed the structure of the planet. These particles contain the
essence of many of the minerals which have been found in the spin
off from these gigantic magnetic vortices. Life then evolved
apparently much as evolved on earth, allowing for different
adaptations to meet the existing environment on Mars during
different epochs in the planet's history.

Compared to my former psychic probes of planets Jupiter
and Mercury, this mental scanning of Mars has been more exciting
and perhaps more significant. My overall impression is that Mars
is still a live planet, which is trying to recover from the

devastating period countless centuries ago when it was peopled
with a civilization equal, if not superior to our own. It is
even possible that some Martians of that day reached our earth
and brought some forms of life here. The feeling comes to me
that there are what I would call molecules of intelligence,
potential shades of life existing throughout the spacious
universe awaiting identification with the right chemical and
electrical elements to become part of limitless living structures.
It is not impossible that our scientific study of biochemical
conditions on Mars will reveal organic evidence of the gases
necessary to produce life.

     Some after impressions: On conclusion of my psychic probe
and while there is still time remaining in the hour assigned for
recording my impressions, I have projected my mind ahead in time
to attempt to foresee the conditions which will exist at the
landing site to which the Viking A is heading. I get the feeling
that the orbiter, when it arrives, before releasing the lander,
will have to continue in orbit until surface conditions have
cleared away because of the violent dust storms and that the
landing may not take place on July 4, 1976 as scheduled. The
second landing site may need to be chosen and there may be
concern even then that the metal feet make upright contact with
the Mars surface. The dust appears to have a reddish cast
lifted by high velocity winds to a height some miles above the
terrain. I have the feeling that some temperature as well as
magnetic changes bring on these storms which die down almost
as quickly as they spring up, yet they could also last for a
number of days. These winds have been responsible for widespread erosion of extensive areas and the friction of the falling
particles has smoothed off some rock formations and lava-caked
regions.

                              END

PSYCHIC PROBE OF MARS II   January 29, 1976

Ingo Swann Response   9 - 9:35 PM EST

Also present: Ed May (EM) and Janet Mitchell (JM)

It looks like that at the North Pole, you can see them
very nicely if you go a little ways outside of earth's outer
atmosphere. The northern hemisphere seems to be pretty much
brilliantly covered with aurora borealis. Very beautiful
tonight. Also, as I go further out, I don't know how we'd ever
find out, but there is a phenomenon which I've never noticed
before, such things as magnetic storms in space. It seems very
easy just to be at Mars but I like to slow the perception down
so we can have an approach to the planet. It seems extremely
red or pinkish, much more brilliant than the last time. I think
I might have to say that there are some dust storms going on
and those are dust particles. Now I can touch, I'm getting deeper
in toward the planet and I can sense or feel the winds. The winds
are blowing. It seems to be a different season than when we were
here before. I mentioned last time the thermal layer of air very
close to the planet's surface and the cold layers are above it.
This time, I suppose it's the winds, there seems to be a different
environment. I would say the winds are strong but not strong,
like I think I read 200 miles per hour. These must be more like
those Santa Ana winds in Los Angeles, which I think go about 70
upwards, somewhere around in there. I think I have come down in
the area of this big volcano. I see some tornados off in the
distance and lightening. I have a sense that the lava shield
around the volcano has a very high mineral content in it, like
lots of iron. Inside the dust cover, the atmosphere cover, the
air color seems reddish. I see two colors of lightening, one
yellow and one greenish. I think that the air is very dry, but
I get a sense of mud also in various places. I don't know how
that could be. OK, let's do the coordinates.

JM: 19.5°N, 34°W

I didn't seem to shift my perspective too much. It seems
near the same place I was. Would you do it again?

JM: 19.5°N, 34°W

I have shifted a little bit to what must be southeast and
this brought me very near or on a plain. Descending to the south
is a series of what looks like cliffs and fractures going south.
I get the impression, what looks to me like the ground at this
point has been swept like by a broom and it seems to be very hard
rock with small cracks in it, like shale. It breaks up in the
wind a little bit. It looks like shale. It looks like reddish
to dark brown in color. It seems to me that if that Viking lands
here, it will land on rock. I see some sort of yellow hills to
the east. It seems twilight at this site. May we go on to the
next one?

      JM: 20°N, 252°W

I have the sensation of skimming over the surface. I
wonder if there could be two kinds of dust. A heavy kind that
is moved by the wind and settles out faster and a finer kind
that is lifted higher and settles out later. I think my attention
got hung up on something, I'm not at that new site yet. I sort of
passed over some sort of hill formations and I had the sense of
some heavier kind of dust, I guess. If I look down where that
might be, it seems to me there is mud and water there. Can I
have the coordinate again?

      JM: 20°N, 252°W

I seem to be on a higher plateau now. Not too much higher,
but higher. At this site I see pebbles,sort of pebbles and something that looks definitely like caked mud, as if there has been
water here and it's all evaporated. There is still some left in
the mud. This looks like a huge basin.of some sort. I forget if
this one is in a basin or not. One of them is in some sort of a
place like that. This mud sort of stands up on ridges. I wonder
if this is like -- as the ground melted, parts of it melted faster
than others and there were some little cave-ins and this created
little small water rivulets and it left little drainways. It looks
up and down like that (gesturing waves). I'd say from about --
little piles from three inches to nine inches or a foot -- but it's
definitely dried out. It's caked.

I was looking at something; I couldn't figure out what it is.
I think those are worms. Small, little tiny worms. Is there such a
thing as sandworms that maybe could live in dry sand or wet mud? I'd
say that's life, wouldn't you? OK, let's do the next one.

JM: 44.3°N, 10°W

I moved immediately to a site that definitely has something
liquid about it, swampish. Can you give the coordinate again?

JM: 44.3°N, 10°W

I'm still hung up on the worms. This one ends me up in
some sort of a flat area with moisture quality about it. There
looks to me to be something that looks like lichens growing all
over it and they are green and gray. I wonder if this could be
something like I have heard that in Northern Canada, I guess,
permafrost melts in the ground and it sort of gives in and becomes   *Caves*
a little soggy here and there and a kind of mold grows immediately,
even on the ice. Of course the ice is in the ground. It's liquid.
I have the feeling that this is sort of on the edge of a receding,
maybe that's the polar cap receding. It's sort of like a lichen-
like mold. I'm pretty sure that it will dry out and blow away
in the wind producing, I guess, billions of spores. I'm struck
by the fact, although I don't see it but I wonder if this is a
type of plant life that goes through one cycle as a mold and
another as a small plant, the kind we talked about before? I
don't know how 4° latitude could make so much difference but there
seems to be a great deal of difference between what is happening
here and what is happening a little further south. Can we go on
to the next one?

JM: 44.2°N, 110°W

I seemed to have passed over a big canyon. May I have that
again, please?

JM: 44.2°N, 110°W

This brings me to a place that looks like it's white with
red dust on it but it's definitely melting. There is a melting
quality here. I can see the -- there is some lightening in the
distance and it gleams in the moisture, I don't know if it's
water or something else. I would be inclined to say that it is
water. Do you suppose it could be spring here? This must be
much higher elevation, maybe that's why there is still ice here.
It looks like ice of some sort to me. Further to the west I see
a tall uplift of some sort which looks black or dark. I'm tempted
to say that I moved to a darker portion, darker side of the planet
and I seem to see in the air some phosphorus glow, it's glowing a

Mars II - Swann Response   4

little bit. Phosphorescent. That's fascinating. I see also
between where I am and what looks like a large escarpment toward
the west, I see something that looks like lava flows, like they've
been rolled up and down like this, like waves of rock. The wind
is singing as it goes through these -- like wind singing around
uplifted rocks. It makes a noise. Somehow I get the impression
of flowers but I don't know why; I don't see them. I'm trying to
figure out if that's water or not. What is heavy water? Is that
water with a different molecular structure or something? Does
carbon dioxide melt and have a liquid form? I think I would say
that it is water but with something different about it. It makes
it a littel more different. But it's definitely moisture of some
sort. Anything else we should look at?

    JM: Can you give us a fuller description of the worms,
like color and size?

    OK, what was that coordinate?

    JM: 20°N, 252°W

    Before, I was tempted to say that these are little tiny
worms that have fur on them. Furry worms. They are whitish-gray.
Maybe some of them are darker like black. I think there must be
several different sizes. It reminds me of that novel, Dune, that
was so good, where the worms eat the mud and take the spores or
whatever is in the mud and digest those and shit out the earth.
I think they are very busy doing something like that. I bet you
they can get frozen and still be alive. Ed, is there anything
that you think I should look at?

    EM: Color impression of the heavy water.

    It seems clear. Maybe with a slight blue glint, bluishness
to it, but it is definitely coming from something white, which I
take it must be the frozen -- whatever it is that is frozen white.
It appears darker as a substance because it has all this dust
melted up with it and makes sort of a muddly flow but I think if
I look at it coming from the ice, it is clear maybe with a blue
touch to it. I think I'd like to stop now.

    JM: May I ask you one more question?

    Yes.

JM: Which sites do you think that they will actually use?

I don't know. That's a prediction.

END

This is after we've stopped and I've had some food but I
think I have to mention that I saw something that scared me a
little bit and that was definitely what looked like electric
lights off in the distance somewhere -- lights.

JM: At which site?

I don't know. It was enough to be shocked. I was a litzle
shocked at that so, as is usual in those unexpected things you know,
the psychic perception retreats almost instantaneously. Amyhow,
they looked suspiciously like electric lights and I think one of
them was kind of reddish. The rest seemed like lights.

PUBLISHER'S NOTE

ngo continues his chapter "9" to include a third remote viewing session, this time done in 1984. To assist in understanding the complex and interwoven narrative regarding the backstory to Mars, SRI, and remote viewing, evident in this section of Ingo's "9," readers may wish to reference *The Stargate Conspiracy: The Truth About Extraterrestrial Life and the Mysteries of Ancient Egypt* by Lynn Picknett and Clive Prince.

In particular, for what follows in Ingo's chapter, pages 123 and 124, from *The Stargate Conspiracy*, helps set the stage:

"In December 1983, [Richard] Hoagland and [Lambert] Dolphin formed the Independent Mars Mission, with $50,000 from SRI's 'President's Fund', an internal funding source under the discretion of SRI's President, Dr. William Miller. Other key people involved in the Independent Mars Mission were Randolpho Pozos (anthropologist), Ren Breck (manager of InfoMedia, the computer conference company run by the thinking person's ufologist, Dr. Jacques Vallée), Merton Davies (a specialist in the cartography of Mars and other planets) and Gene Cordell (a computer-imaging specialist). One of the first to join the new project was physicist John Brandenburg of Sandia Research Laboratories (which specialises in nuclear weapons research)....SRI's connections with the CIA and Defense Department experiments—such as remote viewing—are too well known to be dismissed, and their reputation obviously preceded them. And now they were funding Hoagland's Mars Mission, after having sent Dolphin to Giza in the 1970s...The independent Mars Mission—with its SRI funding and resources—lasted for seven months, until July 1984, when it presented its findings at a conference at the University of Colorado in Boulder.*"
[*See Hoagland's *The Monuments of Mars: A City on the Edge of Forever*.]

Beginning in 1979, a series of events of which I had no knowledge, until 1984, began to take place. I became apprised of these events in a very strange way. In April 1984, I was again at SRI in California, and again immersed in eternal experimentation designed to "perfect" remote viewing "technology." Not associated with our "psychoenergetics" project but familiar with it was Dr. Lambert Dolphin who was part of a project developing ultra-sound and other equipment for the purpose of identifying things underground—among which were ostensibly hidden chambers in the Great Pyramid near Cairo, Egypt.[7]

Lambert telephoned me one afternoon and invited me to come to his office. When I arrived there, he directed my attention to a photograph displayed on a table. "What do you think that is?" he asked. It was obviously a photograph of a pyramid somewhat buried by sand dunes. "Gee," I replied, "is it a new pyramid just discovered in Egypt?"

Lambert avoided answering. Instead "Can you intuit the size of it?" He handed me a magnifying glass, and I studied the structure carefully. "I can't find any points of reference, so I don't know."

"Well," Lambert continued, "I asked you to use your famous intuition, not your logic." I continued, "Oh. Well let's see. I get...large?"

"How large?"

"Well, large."

Then something began to dawn on me: "You mean," I stared at him, "larger than the Great Pyramid?"

Lambert smiled. "Turn the photo over, but you better sit down first." On the back of the photo was a numbered NASA-Viking photograph stamp.

It was a good thing I was sitting down.

Lambert then opened a file folder and pulled out another photograph. "Does this look familiar to you?" I found myself looking at a checkerboard pattern, and up flooded the Mars images of 1975. "Holy shit!" I nearly fainted.

"Is that what you saw in the Mars probe?" Lambert asked.

---

[7] See Moussa, A., Dolphin, L., Mokhtar, G., *Applications of Modern Sensing Techniques to Egyptology*, Menlo Park, CA, SRI International, 1977.

"No, I don't think so. The ones I saw were in a smaller crater. These look like a...well, a whole city. How large is that area?"

"We believe it's about two miles across," Lambert said "Here's another photo of some dome-like structures."

"Wow I've got to call Harold right away."

But we spent the next hour going over all the photos, a number of them, with magnifying glasses. <u>All</u> of them were officially numbered NASA photos.

What had happened began in 1979 when two researchers, Vincent DiPietro and Greg Molenaar, discovered in two NASA photographs (from Viking 1976) of the Martian surface what appeared to be a large carving of a humanoid face lying on the ground and staring outward into space. DiPietro and Molenaar published these photos, and were surprised by the hostile reaction of the planetary science community. The images, this community said, "were simply nothing more than an illusion brought about by light and shadow."

With their integrity now in question, DiPietro and Molenaar devised new methods of increasing the resolution of the digital images which had been taken by the Viking from high above Mars. "The Face" now was more clear. When presenting this enhanced evidence, they were received by even more expressed hostility, and so having spent thousands of dollars over four years, they decided to retreat.

In 1983, though, a science writer and analyst, Richard Hoagland, saw their work. Hoagland acquired more NASA photographs of the surrounding area of "The Face," and became convinced that the entire area possessed geometric "structures" which could not be explained as having occurred naturally. It was a selection of these photos that Lambert Dolphin showed me. The full story of all this, replete with numerous photographs, was first published in 1986 by anthropologist Randolfo Rafael Pozos as *The Face on Mars: Evidence for a Lost Civilization.*

I don't know where I had been during all this, and when Lambert showed me the photos I realized that I had not been paying attention to the gossip networks. The photos, though, now confirmed much of what Harold Sherman and I had psychically "seen" in 1975, and I quickly had copies of them made to send along to Harold, who soon called me almost in tears of relief.

"Heavens," he said, "if anything could prove the existence of psychic abilities, even those to travel to the distant planets this certainly should."

"Well, yes and no, Harold," I replied "Official opinion denies these are artificially-made structures or buildings, even though they are very hard to explain otherwise. I suppose we will have to await a manned landing on Mars when the astronauts can start counting the masonry blocks. If astrophysicists cannot believe their own photos, belief in long-distance psychic travelling is even farther away."

Lambert Dolphin had a proposition for me, which was to "go" to several selected locations on Mars and "see" what was to be "seen" at them. Naturally, I would be happy to do so, but I reminded him that I had now seen the photos, and so naturally could be "expected" to report structures at the sites.

He countered with the idea that interspersed among the selected locations would be sites on the Martian surface which had no apparent "buildings," but that I would not know in advance which was which.

Still, I <u>had</u> seen the photos and since I fully believe in <u>group psychic efforts</u>, I quickly proposed that I would pull a psychic team together. I decided to collect this team on the East Coast where it was unlikely they had seen the limited publication of DiPietro's and Molenaar's "Face." The "target" (the planet Mars) would be presented to them in sealed envelopes and so none of the "psychics" would know what it was before they opened the envelopes at the appointed time.

Thus, on 15 June 1984, came about the First Group Psychic Probe of Specific Martian Locations, which is to say, two years before Randolfo Rafael Pozo's book became widely available in 1986, and three years before Richard Hoagland's more famous book *The Monuments of Mars* appeared in 1987.

And thus, in this way, did remote viewing begin inextricably to become involved with UFO and extraterrestrial matters, a linkage which I certainly did not foresee in 1972, when the first hints of this psychic art were inadvertently rediscovered at the American Society for Psychical Research.

Before going on to describe the group Mars probe, it is perhaps pertinent here to point out that I have not recounted these psychic discoveries to blow my own horn, or claim that either Harold or I were the "first" to discover certain planetary phenomena that science later found out. The descriptions of these "experiments" are necessary in the context of this book in order to illustrate that Earth psychics can penetrate outer space and the far distant planets—penetrating the territory of UFOs and ETs.

Outer space is practically synonymous with extraterrestrials, for it is they who apparently have the wherewithal to travel through it, while our space technologies are yet in their stumbling infancy. In other words, Earth psychics can penetrate the first or principle realm of the extraterrestrials—which is outer space.

Neither Harold Sherman (now deceased) nor I can claim to be the first to discover "buildings" on Mars. In fact, Earthling myths have largely assumed that Mars was once inhabited. Many past psychics have referred to buildings and civilizations of Mars, and earlier astronomers have observed what they assumed to be "man-made" formations, and "lights" on its surface, lights which moved.

The myths of Martians and Martian civilizations clearly and openly flow through modern science fiction stories, and indeed when Earthlings of the past erected possible and variable images of extraterrestrials, they were almost uniformly referred to as "Martians." The human species possesses what might be called "deep genetic memories" of the past, and these include "memories" of Martians and other extraterrestrial phenomena which surface in myths and science fiction tales.

Harold and I front-loaded information about Mars, and this information included the prevailing scientific opinion that Mars had no life, and certainly had no artificially made structures.

But Earth scientists, in fact, did not know this for sure, even though the opinion passed for fact, and was taught to students as fact. In front-loading ourselves with "known" information about Mars, Harold and I wanted to prepare ourselves for better being able to recognize what science did not know or expect about Mars should we run across it. Neither he nor I believed Mars had "buildings" though and had never discussed the possibility. And thus both of us were shocked to "view" things which could only be described as "man-made."

The group psychic penetration of Mars presented certain technical and organizing difficulties all stemming from my firm understanding that group psychics working together at the same time and with the same target generate increases in the psychic power of all of them.

I selected five psychics, of which I was one, but in inviting the others to take part in a joint psychic experiment, and did not specify the nature of it. I indicated that the target would be in the information package they would receive and open only when the experiment began. I indicated that each of them, some in different cities, would need paper, pencils, pens, a tape recorder with several blank tapes available, and about two hours of time during which they would not be disturbed.

But I realized that when the others found out that the target was Mars, they easily could assume they were supposed to find buildings there, for only that could justify interest in doing the experiment in the first place.

To offset this possibility, I asked Lambert Dolphin to provide me five copies of nine selected Martian sites, some of which referred to the "buildings" in the NASA photos, and some of which referred to Martian locations which had no "buildings."

These nine locations were to be specified by Martian geographic co-ordinates only, and be handed to me in batches of sealed envelopes. When I received them, I arbitrarily numbered the batches 1 through 9, and so even I did not know if the sealed coordinates referred to a "building" occupied or unoccupied part of the Martian surface.

After this, much depended on what the five separate psychics "saw" and whether their separate sightings were similar or radically different. If the latter was ultimately the case, then the experiment could be considered a failure.

It was established that the five psychics, with their gathered materials and safe quiet space, would open their packages at 5:30

p.m. EST on 15 June 1984. At that time, they would find out what the target was, and have five minutes to read the experiment's instructions. These instructions then directed them to open the envelope marked #2, which has been selected as the first site by a random method, and by whatever means possible proceed to that specific place on the Martian surface and begin describing and sketching their feelings, images, etc.

Thus, at the appointed time, four other humans besides myself opened their package of materials at 5:30 p.m. (or its equivalent in other cities) and for the first time found that they were supposed to "go" to Mars.

Other than myself, the Mars-bound psychic team consisted of four others whose names I will not specify because the point I am wishing to make here is that they were all <u>humans</u>, which is to say, members of a psychic species which does not generally recognize itself to be psychic, and that what this group did can be done by others members of this species as well. But the four were composed of:

1. A well-known psychic with a long-track record of success;

2. A psychic who had received some training in remote viewing under my auspices;

3. An individual who claimed occasional psychic insights, but had never before participated in any experiment, much less one of the kind at hand; and

4. Another individual who claimed no psychic gifts, but who was a very intelligent and highly placed official in an important intelligence agency (and who participated only after ensuring that anonymity would forever be maintained).

I had included two psychically inexperienced individuals as "controls," to see what would happen to them as a result of participating in a group-psychic enhancing environment.

At the appointed time, then, the four other participants opened their packages and found, as the first cover page:

PROJECT MARS: June/July 1984

After which they had just five minutes to gather their wits, and somehow launch themselves in the red planet's direction.

After the participating subjects had finished their viewing, they were to mail their sketches, notes, and recorded tapes to me. Upon receipt of all the materials, the combined data would be analyzed and compared for similarities and differences. Not until this final step was finished could any of the participants discuss the experiment between them. In this way, then, five ostensible Mars voyagers could hardly sleep for a couple of weeks while awaiting to learn how this experiment turned out.

As it turned out, and with little variation, all the participants described a large crater at whose bottom were "ruins" of buildings, plazas, and "air ventilators" which gave access to a larger underground network or complex—the whole of which gave evidence of "explosions" or "attacks." But most strikingly, on the lip of the crater, high above the "city," was a monumental pyramid-like structure whose specific purpose was difficult to ascertain save that it was hollow inside.

The data provided by all five psychics integrated almost perfectly, and the last step now was to find out if the target site actually referred to one with suspected "structures" on it or to a place on the Martian surface devoid of them. After the combined data was submitted to Lambert Dolphin, we were informed that the site indeed was a suspected "city" within a crater, and that on the crater's rim was a strange, tall structure which cast a long-spiked shadow from the Martian sunshine. At this point, I made a number copies of the combined data of the first group psychic probe of Mars and sent one to each of the participants, copies of which were "leaked" broadly.[8]

Now a very strange and completely unanticipated thing

---

[8] Mars III (Third psychic probe) - Report by Ingo Swann, as found in The Archives of Harold Sherman, Torreyson Library, Archives and Special Collections, University of Central Arkansas, has been reproduced in its original form at the end of this chapter.

occurred which must be pointed out without any idea that doing so consists of sour grapes. I, and others, had naturally assumed that if any correlation occurred between the psychically-acquired information and NASA's Martian photographs, that interest in pursuing this type of penetration of Mars would become heightened. But the exact contrary in fact took place, and a dense blanket of disinterest thereafter ensued. The group Mars experiment demonstrated:

1. That humans are psychic;
2. That even inexperienced psychic humans working in concert with others can provide relatively accurate information;
3. That great distances were no barrier to this type of penetration from Earth;
4. That psychics could assess and relatively describe "buildings" on Mars which many interpret as really existing based on existing photographs;
5. That the group-psychics described the suspected structures in more detail than what is observable via the photographs;
6. If the buildings are in fact there, they were not built by Earthlings, so far as we know but by some extraterrestrial civilization perhaps long gone from Mars; and,
7. That if all the above are even approximately true, then by direct implication humans possess a psychic capability for penetrating UFOs and Extraterrestrials themselves.

In other words, this group-psychic penetration of Mars constituted a breakthrough which independently could be attempted and probably duplicated by many others, and probably even improved upon by them. The overall result, though, was a Silence and Further Disinterest so thick that it could be cut with a knife but to little avail.

This Silence and Disinterest, however, corresponded exactly with the disinterest in psychic potentials I had earlier (and since) encountered among UFO and ET researchers—whose publications and books themselves clearly attest to the fact that telepathic communications between the extraterrestrials and contactee and abductee humans is almost the sole and exclusive mode of that communication.

The implications of this psychic fact are trenchantly avoided by the very investigators who openly include copious mention of it in their published works.

After 1984, it slowly began to dawn on me that something deeper was involved in all this, a deeper something which might, in a causative way, account for the commonly shared but demonstrably unconscious avoidance which ripples time and again through human societies as a whole.

A larger, _much_ larger question eventually loomed into view: why do mass-consciousness humans, as it were, mass-consciously almost "conspire" to avoid certain issues, and consistently so? My investigations into this matter have revealed that _four_ general areas of societal avoidance have existed for quite some time:

1. Sexuality and eroticism.
2. Human psychic phenomena.
3. General societal love.
4. UFOs and Extraterrestrials.

As any reasonably competent researcher will realize, when such disparate areas as those above can be shown to have one or more things in common (in this case, social avoidance, and even social punishment) then there must exist one general explanation which explains all of them.

I postulate that although each of these four categories of social avoidance and punishment can be studied separately amidst vast social protests against so doing, it may be that all four of them have undiscovered "hidden" _psychic_ relationships which are important and applicable to all of them.

It is with good reason I believe, hitherto almost unimagined, that all four of these areas are at least linked with regard to an Extraterrestrial abductee context which is positively awash with sexuality overtones, while the psychic nature of UFO abductee experiences is visible beyond argument.

***

MARS III (Third psychic probe)

JOINT PSYCHIC PROBE OF A SELECTED SITE ON THE
SURFACE OF THE PLANET MARS (five participants),
CONDUCTED ON JUNE 15, 1984.

Background

In March of 1973, an opportunity for dramatically
testing long-distance remote viewing presented itself.
NASA's spacecraft, Pioneer 10 was on its way to the
planet Jupiter, and would begin relaying back data in
December. Could psychics "go" to Jupiter and "bring
back" information that was unknown or unsuspected.

The first outer-space probe was conducted on
April 27, 1973. Two psychics joined together and
attempted to "travel" to the distant planet at the
same time.

A similar probe of the planet Mercury was made
by the same two psychics on March 11, 1974. Mariner 10
began its bypass of Mercury on March 29, and began to
relay back information.

A third and fourth probe took place on June 14,
1975 and January 29, 1976 respectively, to the planet
Mars, prior to the touchdown of two NASA Viking landers
in July of 1976.

The information generated by the two psychics
was analysed and compared with data about the planets
as it became available in the press.

From the start, it was clear that the information
generated by the two psychics differed considerably
with what was known about the conditions on the planets,
and what was expected by opinion and consensus at
the time. Some of the psychically obtained information
does not appear to be correct. Some of this "incorrect"
information must be held in as not proven until future

-2-

more intimate contact with the planets can take place. Some of the psychically-obtained information was proven to be correct, even though at the time it flew in the face of the informed consensus.

Regarding Jupiter, bands like the rings of Saturn were perceived. The existence of such rings was considered imposible up until 1979 when Voyager 2 sent back information during its crossing of the ring system now known to exist around Jupiter.

Regarding Mercury, the unsuspected thin atmosphere was correctly identified by the two psychics in advance of the information coming back from Mariner 10.

With regard to the planet Mars, the two psychics correctly identified the presence of ice and frost, composed of carbon dioxide. The indications of ice found on Mars by Viking 2 surprised scientists. Several other factors of the Martian surface were also described by the two psychics, and later proven correct.

One feature of the psychically-acquired information concerning Mars' that has not yet been proven correct was the presence at various places on the surface of man-made or artifical structures. These were described as dome-like structures, squares or checkerboard patterns, all not of natural origin. Other tall structures comparable to towers that carry high-voltage lines, or beacon structures for aircraft, were also reported. Additionally, artifical satellites, which appeared to be in stationary orbit, were mentioned. These were all in ruins.

During the early part of 1984, a group of scientists at SRI International began a project to acquire hundreds of NASA photographs of the surface of Mars, and subject them to various enhancement techniques. The result of this analysis began to yield features on the surface of Mars for which a natural or geologic explanation was dubious.

One of the psychics that had participated in the earlier planetary psychic probes was invited to gather psychic information about some of these features. He suggested putting together a team of people, each of which would, at the same time, attempt

-3-

to "travel" to the sites indicated (by coordinate only),
and report on the conditions psychically perceived there.

A team of five psychics was gathered. They are
referred to by alphabet numbers:

A - a well-known psychic with a track-record for
long-distance viewing established over forty years ago;

B - an individual was minimal psychic experience and
who has never before attempted a long-distance task;

C - a psychic who has had some training;

D - (anonymous)

E - a well-known psychic who has achieved a reputation
in long-distance remote-viewing capabilities.

The psychics in the group were provided with nothing
more than the Martian coordinates for each of the sites.
As of this writing, only one coordinate has been looked
at. No other information was given in advance. The
psychics were spread throughout the country.

By agreement, the third probe of Mars, with the
first selected coordinate, began at 5.30 p.m., EST,
on June 15, 1984.

-4-

Results

At the start of the viewing session, each viewer
was instructed first to locate Mars. From this initial
contact, general information concerning overall
conditions on Mars was described. Since the coordinate
site was of most interest, we will move directly on to that.

All of the participants first described locating
a crater of large dimensions, and between them they
described a large structure on the crater's lip and
a complex of structures further down on the crater's
floor.

The crater

The crater is of the impact type, but in dimensions
it was seen to be larger than, by comparison, the crater
in Arizona. It contains drifting sand dunes, but also
apparently several different types of debris compared
to being "crunchy, like walking on seashells," or
"pumice-like."

It is "warmer" towards the bottom, with kinds of
debris that appear not to have originated with the original
impact. This debris is described as cinders, hardened
magma or other kind of "melt" that has been solidified.

Smells of warmth, heat were described as "burny,"
"smoke," "ozone-like," "hot," and "erupting," giving
the impression, by post-session analysis, that there
exists in the crater a source or vent of heat that has
also spewed out a limited amount of ejecta.

Moisture was also noted.

The presence of subterranean cavaties, both natural
and non-natural was noted,. in some cases with specifics.

The structure on the lip of the crater

The structure on the lip of the crater was seen
by all participants to be of extraordinary size, compared
to the rest of the crater's environment. It towers above
the lip and the rest of the terrain, and was consistently
described by the team as a "pyramid."

-5-

A. - "...evidence of a past civilization on Mars -
as we have on earth in the pyramids and long before
that."

B. - "It doesn't appear that what I'm seeing is
a natural or physicality of the planet. It's old.
Now I'm getting the impressions of white, looking a
little toward the precipice or whatever it is. It
would appear to be a mountainous surface, a mountainous
kind of structure. I'll draw you a picture of it
because I can't quite, don't quite know what it is."
All I can give you is the shape.

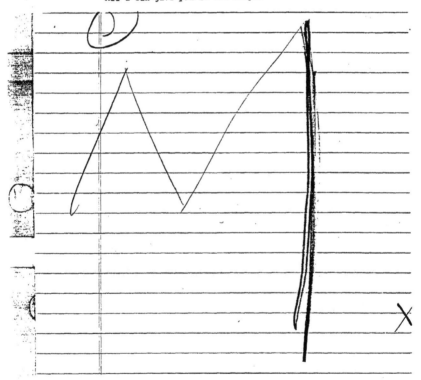

-6-

B. - "...seems to have like a very large pointed
precipice. Almost one that comes straight down. It's
coming in to me rather geometric. Shouldn't be there.
But it is. ... every time I go back to the coordinate,
I go right back to that structure. ... it appears
to be more linear in shape. I think that that's
perhaps what might have been shocking to me is that I'm
getting a very linear, almost geometrical kind of
structure here. There is also something that looks
almost like a platform or something. The construct of
it is of some kind of stone actually, ... a rather stony
kind of structure, rather smooth. It seems a little wind
worn..."

C. - Drew the outline of the towering structure
which was "strange, awesome."

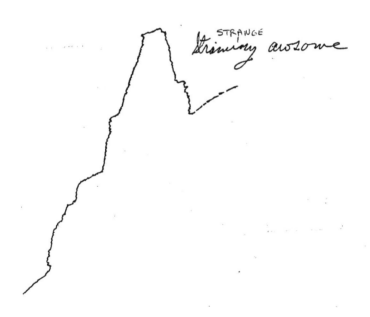

E. - "...in what I'll call the northeast, north,
northeast ... is in the far distance what appears to
be another large but rather organized or sculpted
sloping upward, a quite high edifice. But there ...it
has a lot of sand embanked against it on what appears,
what I would call it, its west side. Near it, here
and there are some quite perpendicular things coming up."

D. - Psychic D made the most lengthy study of
this monumental structure, after he had, as all had,
gotten over his surprise. "If this were on earth, I would
say the feeling was manmade. It has that feeling to it.
But I know this is Mars....I think you can see my dilemma,
my confusion. Okay, I'm going to try it one more time.

"Flat surfaces, patterned, buried, angles, crisp
lines, sloping. Sloping up, down, manmade. A powdery
covering. Pyramid.

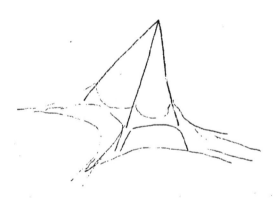

"... specific purpose. It's there for a reason.
Patterned, open interior, chambers, yellowish color.
Very old."

-8-

Participant D made a post-session drawing of the
structure as he conceived it to be.   " I am including
a post-session drawing. This is as detailed as I can
make it, it is a drawing of the pyramid and it's interior
chambers. This pyramid is huge, very, very tall. It had
a very strange feeling about it. When I was in the
chamber, I had the feeling I was in a huge cavern. Yet
it was divided in some way. I didn't see any walls, but
I knew it had different areas or divisions. I have
left them out in my drawing, there is no way I could
draw them in. One thing I did include in the drawing
is the gap or channell separating the room into two
sides (labeled channel). I have drawn only one additional
chamber, although I feel there are several others. I feel
these are all interconnected in some way.

"In chamber one is a pendulum or something suspended.
I perceived movement or activity while in this chamber,
and I feel this activity was coming from this suspended
object. In considering this suspended object I have
a strong feeling that this pyramid was built for,
and still serves a very important specific purpose. This
was not constructed 100,000 years ago and then forgotten.
This may be the big time clock of our solar system or
a huge time capsule or a teleportation station to send
one to the home of its builders. For whatever purpose,
this pyramid is important."

POST SESSION   DRAWING

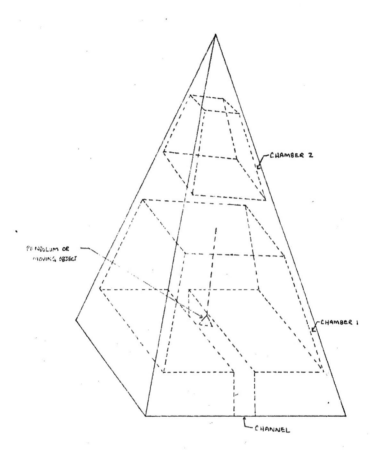

CHAMBER 2

PENDULUM OR
MOVING OBJECT

CHAMBER 1

CHANNEL

-10-

The structures in the crater's interior

The floor of the crater, near or directly beneath the pyramidal structure towering above, appears to contain a grouping of structures that are laid out in a somewhat orderly fashion with larger "avenues" between them. These sit upon or over an underground complex of some type. All of the participents were impressed with the size or out-sizedness of these structures. The nearby location of the source of heat was also noted.

A. - "The first thing I see ... on the left side, is a towering active 'volcano' which seems to be erupting - an outpouring and out-belching of thick, dull, steamy red lava.. In the forefront (of this), a little to the right, is a large dome-like mound, with dark, rock-appearing outlines. Oh! This is terrific. I don't see how this structure can have been formed by nature. Some intelligent creatures must have built it."

C. - "Hollows. Cylindrical. Wide open. Conical, having an inside. Very deep. Like a man-made volcano."

-11-

Sketch made by C.

many over wide area

Comerica
Inside

MAN MADE

Like a ~~mound~~
Volcano

very ancient.

Very deep.

-12-

Each of the participants gave sufficient temperature descriptions to indicate that at the bottom of the crater there exists a thermal vent of some kind, and that it is associated to underground structures. From the descriptions given, it appear that an explosion has taken place, destroying some of the associated underground and surface structures, but leaving others intact or near intact. The source of the ejecta littering the immediate surroundings appears to have come from this vent.

The structures or buildings near this vent were variously described as mounded and non-natural, geometric, linear, huge, interspersed with "avenues" or "plazas" that extend over an area of some large acreage.

C. - Described angular linear lines that intersected that were both "brilliant" and "awesome."

-13-

    E. - Described "avenues" that, in magnitude, were
not disimiliar to those at Teotheuachan, Mexico.

like Teotheuachan

avenues.

like plazas.

big
large.
wind
swept.

-14-

Between these "avenues" or interspersed among them, are squarish structures, and possibly some rounded ones. They are described as being made of stone, but on a monolithic scale.

Sketched by B. —

-15-

Sketched by C. -

hollows
deep
cylindrical
& wide open

E. - Participant E first contacted these structures as flat topped, having empty insides and were in ruins. There were many acres of them, having divisions, and were connected to tunnels or tunneling.

-16-

E. - The structures as progressively sketched:

insides
erupts
ruins

many
acers of them

A flat
topped

divisions

C. - Participant C also began preliminary sketching
of these structures, and indicated their subterranean
connections.

ANgular enclosed
outer soild

Inner open + solid

mean open + solid
down down goes
down

something mechanical

something mechanical
Like elevator

-17-

     E. - Participant E made an effort to engage a
detailed sketch of one of the structures, which in its
size was compared to the Pyramid at Sakara:

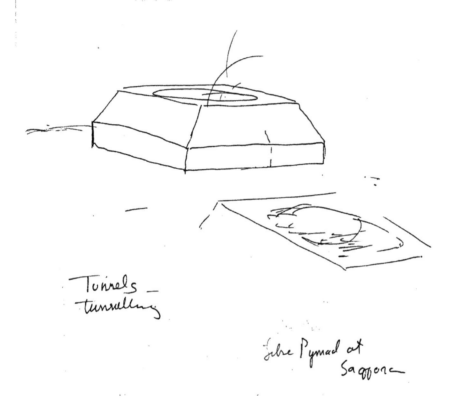

Tunnels -
tunnelling

Jibre Pymad at
Saqqora

-18-

A., C., and E. - All indicated the presence of underground works directly beneath these structures, which are apparently a mix of mounded, rounded and squarish proportions. These structures are indicated (by C and E) to have or be "intakes" of some kind, or perhaps outtakes. They are interspersed by broad platforms or avenues, and are filled with debris, both of both drifting sand or soil as well as debris that is relatable to an explosion of heat intense enough to have brought about lava-like ejecta and have caused melt in some areas.

E. - "I'm in a place where there's quite a bit of sand dunes. Sort of interspersed over rather large ruined structures and pavements. Definitely right-angle walls, made of soft, porous stone. The stones are quite big. This is quite Herculean kind of structures. And they have insides, but I don't see any doors. They are separated by wider avenues, some of which are blocked with sand and some other kind of rubble. Here and there are some quite perpendicular things coming up. Some are tumbled. If I'm to compare them to something that's more familiar to me, I have to say that they're like the air exhaust things that service things like the Holland Tunner. I see twisted metal pylons, bent, very large, and its as if they once had curtains of brick or sheeting on them that's now gone. I have a quite profound feeling about these ruins. These are ruins on a very large expanse, I'm quite in awe."

The underground cavaties were not too much commented upon by the viewers, one of which was reluctant, and so stated, of going into them.

A. - "I am looking at the large mound near the erupting area. I feel that excavations in and around this area may reveal undergound caverns, possibly hieroglyphic letterings or carvings..."

C. - "Like a subway, deserted, cold...chills."

D. - "Strange temperature. Feel hot, but not on the surface.

E. - "Echoes here. Noises. And I can't comprehend how that can be. Things underground here. Corridors and rooms. Some are empty, some have debris of some kind. Large, windswept tunnels and tunnelling."

-19-

Preliminary assessment

After the viewings were completed, and all the
original materials were in the hands of third parties,
the organizer of the psychic team was shown an enhanced
photograph of the coordinate site. It reveals a large
crater with an anomalous pyramid-like structure on its
rim. This feature is anomalyous due to the fact that
if it had been a natural formation it would not have
withstood the original impact that created the crater
in the first place. It might be an upthrust of some kind,
but it throws a shadow across the landscape that clearly
delineates its non-natural pyramidal form.

In front of this structure, on the crater side
appears to be an additional "platform," which can
not be clearly identified. Beneath the pyramid, in
the crater, are a few hints of an organized formation.

All of the participants in the psychic team found
a pyramidal feature on the lip of the crater, and correctly
identified that the site, indicated only by coordinate,
was in fact a crater.

All of the viewers were surprised in discovering
that they felt the structures viewed were of a non-natural
origin, and in most cases this surprise was empathetically
put.

The heat-vent or man-made volcano indicated by most
of the viewers does not show on the photo-enhanced evidence.
Yet all the factors combined between the five psychics
leads to the conclusion, based on their evidence, that
in the bottom of the crater, or near the bottom, there
exist additional structures of some majestic proportions,
and that these are in some fashion connected to the
purpose of the complex, which has, itself, been abandoned
and has, apparently, suffered from some eruption of
a kind that has strewn the vicinity with debris, yet has
not demolished all the complex itself.

The organizer of the psychic team asked an independent
third party highly trained in advanced technologies (who
wishes, obviously, to remain anonymous) to make a best guess
based upon the psychically-obtained information. His
best guess is here quoted in full:

-20-

"General observation: Technological evolution
is measured, in good part, by increasingly greater energy
requirements. Many of the most advanced tools of science,
the wellspring of technological change, demands ever larger
power supplies. This is especially true of lasers,
masons and particle accelerators.

"Data and analysis: The RV data suggest that the
floor of the large crater is situated in the proximity
of subsurface volcanic activity (i.e. a dormant
volcano). The settlement might indeed have been
established at this site solely to exploit this
geophysical phenomenon; so that the inhabitants could
harness the virtually unlimited thermal energy present
in the liquid magma.

"Three possible reasons for finding the site in
its present condition:

"1. An unpredicted/uncontrolled eruption, or
frequent periodic, of the subsurface volcanic strata.

"2. An accident involving the power station
(conversion plant) allowing ejecta to explode up through
the deep conduits (ventillating shafts?) resulting in,
at least, irreparable damage to the power station.

"3. A terrestrial catastrophe (sudden or prolonged)
forcing evacuation of the settlement and leaving the
power station unattended. This resulted in an eventual
breakdown (meltdown?) and led to the description of
the events described in 2. above."

***

AFTERWORD

BY: THOMAS M. MCNEAR,
LIEUTENANT COLONEL U.S.
ARMY (RET.)[9]

As a young Army Captain I was assigned to the Army's Stargate "psychic spying program" from 1981-1985. It was during this assignment that I met the amazing Mr. Ingo Swann. I was the first Army officer Ingo trained in remote viewing, and the only member he trained in all stages of remote viewing.

It was 1981. My friend Robert Cowart and I were Captains sitting in class at the Military Intelligence Officers Course (MIAOC) at Fort Huachuca, Arizona. As the day was winding down, in came three personnel with what was described to us as a "psychological survey." The Army was trying to determine the psychological profile of the average intelligence officer—or so we were told. They said we didn't even need to put our names on the paper; it was just an anonymous survey. We all completed the surveys and went home for the day. Unbeknownst to us, as we turned in our surveys, they were writing our names on them.

The next day, Rob and I learned the rest of the story. The "survey" questions were actually intended to identify 14 traits the Army believed indicated a predisposition to being a good psychic. Rob and I had both scored very well. Rob hit 11 of the 14 categories and I had scored 14 of 14.

We were briefed on a highly classified intelligence collection program called Grill Flame—later the name was changed to Stargate. We were asked if we would be willing to accept an assignment to Fort Meade, Maryland to become "psychic spies?" Who could turn down an offer like that? We both accepted.

[9] Thomas McNear's brief biography is provided in the Selected Bibliography section.

We finished MIAOC and reported to Fort Meade. There we learned that we were going to be trained in a psychic skill known as remote viewing; in fact, we were going to be trained by the father of remote viewing himself, Mr. Ingo Swann.

In early 1982 Rob and I flew to Menlo Park, California, to Stanford Research Institute International (SRI-International) to meet Mr. Swann. We didn't know what to expect. Rob and I admitted to each other we were a bit nervous about that first meeting. Would he be "weird?" Could he read our minds? He had a security clearance, but what could we discuss with him? Would we get along?

Years later, Ingo admitted to me that he was "scared shitless" during our first meeting. He said he didn't know what to expect from career Army officers...would we be weird; would we get along?

Rob and I trained with Ingo at SRI three times, typically for two-week intervals. After the third such training, Rob was diagnosed with cancer and was medically retired from the Army. That was a shock; I missed my friend and training partner. For the next three and a half years it was just Ingo and me; for the first three years at SRI in California, and for the final six months in the SRI offices in New York City. Toward the end of my training, Ingo trained four more Army personnel in stages 1-3 before Ingo ended his contract with the Army.

During my almost four years working with Ingo, I came to recognize him as much more than just the father of remote viewing; Ingo was a creative genius, a gifted artist, a writer, a teacher, a mentor, a visionary...an inter-galactic psychic time-traveler...and a friend. Our friendship lasted until his passing in 2013. I, and many others who knew him well, miss him greatly.

Ingo was an excellent trainer and mentor. He saw our training as his way to pass on what he knew to subsequent generations of remote viewers seeking what he called a psychic renaissance. He could be demanding, ensuring his students understood how the psychic signals presented themselves and how to avoid allowing our ever-present analysis to interfere. He was demanding because this was important to the intelligence community and to him; this was his legacy.

***

In the summer of 1984 Ingo asked me to participate in a special remote viewing. He said this would be a team effort. Five of us, Ingo, myself and three others, would remote view the same target at the same time. On 15 Jun 1984, as Ingo addresses in "9, The Psychic Probes of Mars," the five of us remote viewed Mars.

My participation in this viewing of Mars and what followed is very interesting. I was one of the five (probably member number 2, but possibly number 4) and on 15 Jun 1984 at 1730 EST we, collectively, went to Mars. My experiences there were profound. As Ingo stated, we all quickly packaged up our results and mailed them to him. Because Ingo was in a hurry, I failed to make copies of any of the materials thinking I could get copies from him later should I need them.

While visiting Mars, I encountered entities and communed with them. The fact that they were there, and what they told me, was profound, yet so inconceivable to me that I filed it away in the dark recesses of my mind trying to not make too much of it. Perhaps I was mistaken. Though this encounter haunted me through the years, I was able to put the encounter back into my mental box because to seriously consider that I had communed with multiple entities on Mars to me seemed absurd - inconceivable.

Then in 2014, a year after Ingo's passing, I was reading Joe McMoneagle's book *Mind Trek*. In it, Joe recounts his remote viewing of Mars on 22 May 1984—24 days before Ingo's five-person team.[10] Joe speaks of the same entities, the same profound sense of loss, the same structures, the same inner chambers...to this day I am reluctant to discuss it, but if you want to know more, read *Mind Trek*.

As I was reading, I jumped up from my seat and began pacing the floor. All of my memories leaped from the box in which I had stored them...I had to get a copy of my materials, but from whom; Ingo had passed the year before.

---

[10] See The Central Intelligence Agency's Mars Exploration, document RDP96-00788R00190076001-9:
https://www.cia.gov/library/readingroom/docs/CIA-RDP96-00788R001900760001-9.pdf

Since reading Joe's book I have been unable to ignore what I experienced, yet I have been unable to really discuss the experience with anyone because I know what I know in that "vague-ish, remote viewing sort of knowing." If I were to discuss it, questions would arise that I am unable or unwilling to answer. Logic would attempt to fill in the blanks and my true recollections would be lost. I recently read, "To know what you want to share with the world because something has been transmitted to you is to know loneliness and real solitude, because no one else can share it." That hit me; I knew exactly what it meant.

I have subsequently tried to get the information from the University of West Georgia – the home of Ingo's archives.[11] They have searched Ingo's archives, but so far have been unable to locate the information I recall mailing to Ingo. The one thing I most remember of the packet was a drawing, not a sketch, but a drawing of a hollow pyramidal structure. I knew it was so important that rather than simply sketching it, I got out a ruler and paper and did the best I could to draw what I had discovered. The phrase I used to describe this object has come to me hundreds of times since; I said I don't know what it is or its purpose, "but I know when we find it we will know..." we will know what it is, why it is there and who left it.

Regarding the five, I only know that it was Ingo, I, and three others. As he said in "9," he made a vow to keep our identities secret. Perhaps others had requested that as well, but this was my request to Ingo. Because I was continuing my career in the intelligence community and still had a security clearance, I felt I needed to maintain my anonymity.

What, or who did we discover on Mars? I don't believe we will know that in our lifetimes. But if you want to know more, read Joe McMoneagle's *Mind Trek*, or better yet, relax, clear your mind, and visit Mars for yourselves. I would love to hear your experiences.

Send your experiences to:
RemoteViewMars@gmail.com

---

[11] See the notes on the University of West Georgia in the Selected Bibliography section.

S ubsequent to the first edition, Thomas (Tom) McNear and researcher Brian Lawlan undertook an investigation into the identities of the five remote viewers. What follows are their notes:

Although descriptions used in "9" and in "Mars III" to identify the 1984 remote viewers are similar, they are at the same time different, allowing for some confusion. The confusion is due to the fact that Ingo used *numbers* in "9" and *letters* in "Mars III" and did not make a one-to-one association between the two. True to Ingo's commitment for anonymity, he scrambled the correlation of the letters and numbers identifying the psychics.

From Ingo's own notes and notes found in Harold Sherman's archives, we were able to establish that Psychic 1 from "9" was Harold Sherman. We were also able to verify using Tom's own notes that Tom was Psychic 2. Even though "9" did not list the fifth psychic, it is clear this was Ingo himself. Ingo quoted Harold Sherman's own account of his (Sherman's) viewing, thereby clearly identifying Sherman as viewer A in "Mars III." We also established that Psychic E was Ingo.

Thus, our first undertaking was to confirm Tom's identification within the "Mars III" document. The descriptions of both 4 and D include "anonymity," implying they are the same, but they are not. We started with what was known: Tom McNear was viewer number 2. This was confirmed by his notes he found from 15 June 1984.

The identities of Psychics C and D in "Mars III" have been confirmed by comparing handwriting samples with known samples of the suspected participants. We were able to determine Psychic D was Tom McNear and Psychic C was a person known to Tom. Because Psychic C has passed away, we have respected Ingo's promise of anonymity; Psychic C will be referred to as "Unnamed." This provides us with certainty regarding the identities of Psychics A, C, D and E from the "Mars III" document. Only psychic B remains unknown.

Now, knowing four of the five psychics (A, C, D and E) in "Mars III," we applied that knowledge to Psychics 1-4 and 5 in "9." The relative ambiguity of the descriptions of Psychics 3 and 4 required much examination. Unnamed must be either Psychic 3 or 4 because the identities of the others were unambiguously known. Because Psychic 4 was a "highly placed official in an important intelligence agency" and Tom knew Unnamed was not such a person, Unnamed could only have been Psychic 3. Only Psychic 4 remains unknown. Because of the above, we can, state the following:

- Harold Sherman was psychic 1 and A.
- Tom McNear was psychic 2 and D.
- Unnamed was psychic 3 and C.
- Ingo Swann was psychic 5 and E.

Regarding the unknown psychic, we have a suspicion as to the identity of this individual but currently have no confirmation. "Mars III" contains two sketches from Psychic B but no handwriting for analysis. In "Mars III" Ingo quoted two paragraphs from Psychic B's session. Several words/phrases contained in these paragraphs may help to identify Psychic B, but one would likely have to know Psychic B to make that association. Because remote viewers are "speaking" from their right-hemispheres, making that association may be more difficult.

Below is a table associating Ingo's descriptions with the remote viewers:

Chapter "9"

Number	Description	Psychic
1	A well-known psychic with a long-track record of success.	Harold Sherman
2	A psychic who had received some training in remote viewing under my auspices.	Tom McNear
3	An individual who claimed occasional psychic insights, but had never before participated in any experiment, much less one of the kind at hand.	Known but not being disclosed as the person has since died
4	Another individual who claimed no psychic gifts, but who was a very intelligent and highly placed official in an important intelligence agency (and who participated only after ensuring that anonymity would forever be maintained).	Unknown
5	No Description (Viewer number five was not listed in Penetration SE).	Ingo Swann

"Mars III"

Letter	Description	Psychic
A	A well-known psychic with a track-record for long-distance viewing established over forty years ago.	Harold Sherman
B	An individual was (sic) minimal psychic experience and who has never before attempted a long-distance task.	Unknown
C	A psychic who has had some training.	Known but not being disclosed as the person has since died
D	(Anonymous)	Tom McNear
E	A well-known psychic who has achieved a reputation in long-distance remote-viewing capabilities.	Ingo Swann

SELECTED BIBLIOGRAPHY

NOTES

The topic of Moon anomalies (including convincing evidence for artificial structures) is complicated by coverup agendas that many do not care to infringe upon. Thus, the topic is shunted aside from larger sectors of inquiry—such as science, space studies, conventional lunar studies, mainstream media, and Ufology.

Even so, certain unofficial sources are replete with well-interpreted evidence, documentation, and quite excellent bibliographies that can act as guide to more extensive information. These unofficial sources are indicated by an asterisk (*).

With regard to telepathy, hardly any sources have addressed it in other than superficial ways. Although many abductees have indicated that extraterrestrials communicate via some telepathic form that is non-language dependent, I have decided not to include references to the abductee literature which is quite large and easily available.

A large and vivid vacuum of information exists regarding the phenomena of group mind and subliminal group consciousness management that might be invasively influenced by various means such as forms of super-telepathy as yet unacknowledged as existing.

Selected Internet addresses regarding UFOs and unusual lunar phenomena as well as books on the ancient life of Mars and the links between SRI and Giza have been introduced into this bibliography.

BIBLIOGRAPHY

*Andrews, George C., *Extra-Terrestrial Friends and Foes.* (Lilburn, Georgia: IllumiNet Press, 1993).

Berelson, Bernard, and Steiner, G.A., *Human Behavior: An Inventory of Scientific Findings.* (New York: Harcourt, Brace, and World, 1964).

*Bergquist, N. O., *The Moon Puzzle.* (Copenhagen: Grafisk Forlag, 1954).

Berliner, Don, with Marie Galbraith and Antonio Huneeus, *Unidentified Flying Objects Briefing Document: The Best Available Evidence* (A limited publication presented by CUFOS, FUROR, MUFON, December 1995).

Blum, Howard, *Out There.* (New York: Simon & Schuster, 1990).

Bourret, Jean-Claude, *The Crack in the Universe: What You Have Not Been Told About Flying Saucers.* (Suffolk, England: Neville Spearman, 1974).

Brandenburg, John. *Life and Death on Mars: The New Mars Synthesis.* (Kempton, Illinois: Adventures Unlimited Press, 2010).

*Brian, William, *Moongate: Suppressed Findings of the US Space Program.* (Portland, Oregon: Future Science Publishing Co., 1982).

Chatelain, Maurice, *Our Ancestors Came From Outer Space.* (New York: Doubleday, 1978).

Cherrington, Ernest H., *Exploring the Moon Through Binoculars & Small Telescopes.* (New York: Dover Publications, 1969).

*Childress, David Hatcher, *Extra-Terrestrial Archaeology.* (Stelle, Illinois: Adventures Unlimited Press, 1994).

Clark, Jerome, *The UFO Encyclopedia* [in three volumes]. (Detroit: Apogee Books, 1990).

CNI News (Global News on Contact with Non-Human Intelligence). Web address: http://cninews.com and http://www. iscni.com/

Corliss, William, *The Moon and the Planets: A Catalogue of Astronomical Anomalies.* (Glen Arm, Maryland: The Sourcebook Project, 1985).

*Cornet, Bruce, "Memorandum on Unusual Lunar Features" in *CE CHRONICALS,* July-August 1994 Issue on *Lunar Anomalies* (10878 Westheimer, Suite 293, Houston, Texas 77042).

DiPietro, V. and Molenaar, G., *Unusual Martian Surface Features.* (Glenn Dale, MD: Mars Research, 1982).

Dixon, Norman F., *Subliminal Perception: The Nature of a Controversy.* (London: McGraw-Hill, 1971).

Fawcett, Lawrence and Greenwood, Barry, *Clear Intent: The Government Cover-up of the UFO Experience.* (Englewood *Cliffs,* New Jersey: Prentice Hall, 1984). Also published as *The* UFO *Cover-Up.* (New York: Simon & Schuster, 1992).

*Firsoff, V.A., *Strange World of the Moon.* (New York: Basic Books, 1959).

Garrett, E. J., *Telepathy.* (New York: Creative Age Press, 1941).

Good, Timothy, *Above Top Secret.* (New York, William Morrow, 1988). *Alien Contact.* (New York: William Morrow, 1993).

*Guiley, Rosemary Ellen, *Moonscapes.* (Englewood Cliffs, New Jersey: Prentice Hall, 1991).

Hamilton, William F., *Cosmic Top Secret: America Secret UFO Program.* (New Brunswick, New Jersey: Inner Light Publications, 1991).

Hancock, Graham and Bauval, Robert, *The Message of the Sphinx: A Quest for the Hidden Legacy of Mankind.* (New York: Crown Publishing Group, 1996).

Hill, Harold, *A Portfolio of Lunar Drawings.* (New York: Cambridge University Press, 1991).

Hoagland, Richard. *The Monuments of Mars: A City on the Edge of Forever.* (Berkeley, California: North Atlantic Books, 1987).

Hyslop, James, *Contact with the Other World: The Latest Evidence as to Communication with the Dead.* (New York: Century, Co., 1919).

Jessup, Morris K. *The Expanding Case for the UFO.* (New York: Citadel Press, 1957).

Key, Wilson Bryan, *Subliminal Seduction.* (New York: New American Library, 1973).

Knapp, George, *UFOs: The Best Evidence.* (UFO Audio-Video Clearing House, P.O. Box 342, Yucaipa, California 92399).

*Kono, Kenichi, *The Moon Has Structures.* (Tokyo: Tama Publisher, 1980) (Note: This book has NOT been translated into English but contains some astounding photographs).

*Leonard, George, *Somebody Else Is On The Moon.* (New York: Pocket Books, 1975).

Lowell, Percival. *Mars and its Canals.* (New York: Macmillan Co., 1906).

*Marrs, Jim, *Alien Agenda.* (New York: HarperCollins Publishers, 1997).

McGinnis, Paul, *McGinnis Military Secrecy.* Web address: http://www.frogi.org/secrecy.html

McMoneagle, Joseph. *Mind Trek: Exploring Consciousness, Time, and Space Through Remote Viewing.* (Charlottesville, Virginia: Hampton Roads Publishing, 1993/1997.

*Middlehurst, Barbara M., et al, *Chronological Catalog of Reported Lunar Events.* (NASA Technical Report R-227, 1968).

Moussa, Ali Helmi, Lambert, Dolphin, and Mokhtar, Gamal, *Applications of Modern Sensing Techniques to Egyptology.* (Menlo Park, California: SRI International, 1977).

Packard, Vance, *The Hidden Persuaders.* (New York: David McKay, 1957).

Picknett, Lynn and Prince, Clive. *The Stargate Conspiracy: The Truth About Extraterrestrial Life and the Mysteries of Ancient Egypt.* (New York: Berkley Publishing Group, 2001).

Pozos, Randolfo Rafael. *The Face on Mars: Evidence for a Lost Civilization.* (Chicago: Chicago Review Press, 1986).

Randles, Jenny, *Alien Contact: The First Fifty Years.* (New York: Barnes & Noble, 1997).

Rense, Jeff. Computer Enhancements by Liz Edwards. "Astonishing Intelligent Artifacts Found on Mysterious Far Side of the Moon." Sightings Website.

Ross, Daniel K., *UFO's and the Complete Evidence from Space.* (Walnut Creek, California: Pintado Publishing, 1987).

Sagan, Carl, Ed., *Communication with Extraterrestrial Intelligence.* (Boston: M.I.T. Press, 1963).

Schiaparelli, G. V. *Il pianeta Marte.* (Milano, Italy: Vallardi, 1893 and 1909).

Schiaparelli, G. V. *La vita sul pianeta Marte.* (Milano, Italy: Vallardi, 1895).

Shklovskii, L. S. and Sagan, Carl. *Intelligent Life in the Universe.* (San Francisco: Holden-Day, Inc., 1966).

Stacy, Denis, "Cosmic Conspiracy: Six Decades of Government UFO Cover-ups." (*OMNI Magazine:* In a six-part series, beginning in Vol. 16, No. 7, April 1994).

*Steckling, Fred, *We Discovered Alien Bases On The Moon.* (Los Angeles: GAF Publishers, 1981).

*Sullivan, Walter, *We Are Not Alone.* (New York: McGraw-Hill, 1964).

Taylor, Eldon, *Subliminal Communication.* (Salt Lake City: JAR, 1988).

Thompson, Richard L, *Alien Identities: Ancient Insights into Modern UFO Phenomena.* (San Diego: Govardhan Hill Publishing, 1993).

Ubell, Earl. "The Moon Is More Of a Mystery Than Ever." New York Times Magazine, New York, April 16, 1972, 32.

UFO NET. Anthony Chippendale, Ed. Web address: http:// www.ufo-net.clara.net [URL inactive].

UFO ROUNDUP. Joseph Trainor, Ed. Web addresses: http://ufoinfo.com/roundup.

"UFOs on Air Defense Radars." Rabochaya Tribuna, Moscow, April 19, 1990.

United States. Central Intelligence Agency. *Mars Exploration,* May 22, 1984, https://www.cia.gov/library/readingroom/docs/CIA-RDP96-00788R001900760001-9.pdf.

Vallée, Jacques, *Dimensions: A Casebook of Alien Contact.* (New York: Ballantine, 1989).

Weiner, Tim, *Blank Check: The Pentagon's Black Budget.* (New York: Warner, 1990).

Wilkins, Percival H., *Our Moon.* (London: Frederick Muller, 1954).

*Wilson, Don, *Our Mysterious Spaceship Moon.* (New York: Dell, 1975).

Wright, Susan, *UFO Headquarters: Investigations on Current Extraterrestrial Activity.* (New York: St. Martin's Press, 1998).

UNIVERSITY OF WEST GEORGIA

Special Collections, serving as the repository for rare manuscripts, books, films, photographs, sound recordings, and other formats in a number of specialized areas, within the Irvine Sullivan Ingram Library at the University of West Georgia in Carrollton, Georgia, USA, holds Ingo's archives as well as those of Dr. Mitchell and Dr. Krippner.

Ingo's notes and materials regarding his planetary remote viewing sessions are found within his archives as part of **Series 6: Research and Manuscripts, Planetary Remote Viewing (RV) Files**:

March 11, 1974 – Mercury RV – An Experimental Remote Viewing Probe of a distant Planet: J. Mitchell, H. Sherman and I. Swann, New York and Mountain View, AR.

March 18, 1975 – Moon RV – I. Swann.

June 14, 1975 – Mars RV 1 – I. Swann and H. Sherman

January 29, 1976 – Mars RV 2 – I. Swann and H. Sherman Swann and Sherman independently probed Mars from their respective locations. With Ed May and Janet Mitchell present. Transcript of Swann and Sherman's verbalizations.

June 15, 1984 – Mars RV 3 – I. Swann and four others.

UNIVERSITY OF CENTRAL ARKANSAS

The University of Central Arkansas Archives is dedicated to the acquisition and preservation of historical documents that pertain to the state of Arkansas. The holdings of the UCA Archives are very diverse and include Military, Environmental, Mass Communication, Art, Religion, Sports, Education, Political History, and many other important topics., including the archives of Harold Sherman.

Harold Sherman's and Ingo Swann's notes and materials regarding their planetary remote viewing sessions are found within Harold Sherman's archives as :

June 14, 1975 – Psychic Probe of Mars I: Harold Sherman and Ingo Swann

January 29, 1976 – Psychic Probe of Mars II: Harold Sherman and Ingo Swann.

June 15, 1984 – Psychic Probe of Mars III – Ingo Swann Report

DR. STANLEY KRIPPNER

Stanley Krippner, Ph.D. was the Alan Watts Professor of Psychology at Saybrook University for nearly 50 years. He is a Fellow in five American Psychological Association (APA) divisions, and past-president of two divisions. Formerly, he was director of the Kent State University Child Study Center, Kent Ohio, and the Maimonides Medical Center Dream Research Laboratory, Brooklyn New York.

He is co-author of *Extraordinary Dreams, Personal Mythology, Dream Telepathy, Sex and Love in the 21st Century, The Voice of Rolling Thunder, Demystifying Shamans and Their World, A Psychiatrist in Paradise: Treating Mental Illness in Bali, Post-Traumatic Stress Disorder: Biography of a Disease, Dreamworking: How to Use Your Dreams for Creative Problem-Solving,* and *Haunted by Combat: Understanding PTSD in War Veterans,* and co-editor of *Healing Tales, Healing Stories, Mysterious Minds, Debating Psychic Experience, The Psychological Impact of War on Civilians: An International Perspective, Dreamscaping: New and Creative Ways to Work with Your Dreams, Broken Images, Broken Selves: Dissociative Narratives in Clinical Practice,* and *Varieties of Anomalous Experience: Examining the Scientific Evidence.*

He edited or co-edited ten volumes of *Advances in Parapsychological Research.*

He received the APA Award for Distinguished Contributions to the International Development of Psychology in 2002, the Society for Psychological Hypnosis Award for Distinguished Contributions to Professional Hypnosis in 2002, the Ashley Montagu Peace Award in 2003, and lifetime achievement awards from the International Association for the Study of Dreams, The International Network on Personal Meaning, the Society for Humanistic Psychology, the Parapsychological Association, and the J.B. Rhine Lifetime Achievement in Parapsychology Award .He received that Award for Distinguished Contributions to Professional Hypnosis from the Society for Psychological Hypnosis, the Senior Contributor Award from the Society for Counseling Psychology, the Pathfinder Award from the Association for Humanistic Psychology, and the Charlotte and Karl Buhler Award from the Society for Humanistic Psychology.

In addition to his APA listings, he holds Fellow status in several other organizations including, the Society for the Scientific Study of Religion, the Society for the Scientific Study of Sexuality, the American Educational Research Association, the Western Psychological Association, the Society of Clinical and Experimental Hypnosis, and the American Society of Clinical Hypnosis. He is a Founding Fellow of the Association for Psychological Science and the American Academy of Clinical Sexologists. He holds Diplomate status in the American Board of Sexology and the International Academy of Behavioral Medicine, Counseling, and Psychotherapy. He is certified as an Advanced Alcohol and Drug Counselor.

Dr. Krippner has conducted workshops and seminars on dreams and/or hypnosis in Argentina, Brazil, Canada, China, Colombia, Cuba, Cyprus, Ecuador, Finland, France, Germany, Great Britain, Italy, Japan, Lithuania, Mexico, the Netherlands, Panama, the Philippines, Portugal, Puerto Rico, Russia, South Africa, Spain, Sweden, Venezuela, and at four congresses of the Interamerican Psychological Association. He is a member of the editorial board for the *Revista Argentina de Psicologia Paranormal*, and the advisory board for International School for Psychotherapy, Counseling, and Group Leadership (St. Petersburg) and the Czech Unitaria (Prague). He holds faculty appointments at the Universidade Holistica Internacional (Brasilia), El Universidad Transpersonal (Puebla, Mexico), and the Instituto de Medicina y Tecnologia Avanzada de la Conducta (Ciudad Juarez). He has given invited addresses for the Chinese Academy of Sciences, the Russian Academy of Pedagogical Sciences, and the School for Diplomatic Studies, Montevideo, Uruguay. In 1959, he received the Service to Youth Award from the Young Men's Christian Association in Richmond, Virginia, his first award.

DR. JANET LEE MITCHELL

Janet Lee Mitchell received her Ph.D. in experimental cognition from City College of the City University of New York. Dr. Mitchell has had a close association with and commitment to the goals of the American Society Psychical Research (ASPR) for nearly five decades. During this time, she has designed, conducted and published research, as well as helped to raise funds to support the ASPR's research program. Dr. Mitchell ran many ASPR events for the research community and the general public in order to help further the understanding of evolving human abilities. From 1980-1982, Dr. Mitchell was Chairperson of the ASPR Voting Members' Steering Committee. She worked as an ASPR Research Associate with Dr. Karlis Osis from 1967 to 1974 and in 1978.

Dr. Mitchell's research includes ESP and changed states of consciousness, electrophysiological and biofeedback techniques, and research on psychokinesis and on out-of-body experiences with gifted subjects. Her dissertation research on cognitive styles and ESP was conducted at the ASPR, with Dr. Gertrude Schmeidler as program advisor. Dr. Mitchell was awarded an ASPR Graduate Scholarship and a City College Research Fellowship, as well as research grants for her experimental work.

She has published numerous research articles and books. She has lectured internationally and continues to devote an increasing amount of time to helping the general public understand more about psychical research.

Dr. Mitchell has demonstrated a deep commitment to the field of parapsychology and, in particular, to the ASPR. She supports an interdisciplinary approach to psychical research to further the understanding of the phenomena and the acceptance of parapsychology within the broader scientific community.

She is the author of:

Out-of-Body Experiences: A Handbook. (Jefferson, NC: MacFarland Publishing, 1981).

Conscious Evolution, (New York: Ballantine Books, 1990).

Her articles include:

Out-of-the-Body Vision. Psychic, March/April 1973, 44-47.

Out-of-Body Experiences and Autoscopy. The Osteopathic Physician, April 1974, 44-49.

A Psychic Probe of the Planet Mercury. Psychic, June 1975, 16-21.

Further Investigation of PK with Temperature Records. Research in Parapsychology 1974, Metuchen, NJ: Scarecrow Press, 1975, 71-73.

The Astral Journey (Book review). Psychic, October 1975, 46.

PK Effects on Temperature Recordings: An Attempted Replication and Extension. Research in Parapsychology 1975, Metuchen, NJ: Scarecrow Press, 1976, 67-69.

Dream Reality (Book review). Psychic, November 1976, 46.

A Table to Measure Levitation and to Control for Normal Pressure. Journal of the American Society for Psychical Research, 1977, 71, 51-53.

Physiological Correlates of Reported Out-of-Body Experiences. Journal of the Society for Psychical Research, 1977, 49, 525-536.

Learning to Use Extrasensory Perception (Book review). New Realities, July 1977, 28-29.

Out-of-Body Vision. In Rogo, D. S. (ed.), Mind Beyond the Body, NY: Penguin Books, 1978, 154-161.

Psychic Space Exploration. Fate, December 1979, 77-81.

Women in Parapsychology. Research in Parapsychology 1978, Metuchen, NJ: Scarecrow Press, 1979, 25-29.

Dreams and ESP: Review of the Experimental Literature. New England Journal of Parapsychology (in press).

THOMAS MCNEAR

Thomas (Tom) McNear, Lieutenant Colonel, U.S. Army (Ret.) was an original member of the Army psychic spying program known today as Stargate. He was the first member of the program to be personally trained in Coordinate Remote Viewing (CRV) by Ingo Swann and the only member Mr. Swann trained through all stages of remote viewing. He was the "proof of principle guinea pig."

Tom was considered by Ingo Swann to be his best student ever. A fellow remote viewer wrote: "Tom's results were not just impressive. Some could even be considered spectacular." In 1985 Tom wrote the first CRV manual based on his training with Mr. Swann.

After serving in the Army's remote viewing program from 1981-1985, Tom continued a successful career in Army counterintelligence and counterespionage. He retired from active duty in 1997 and continued to serve the Army as a civilian intelligence officer until January 2019. Tom has a Master's Degree in Counseling Psychology from Saint Mary's University in San Antonio, Texas.

In 1984, Tom joined Ingo Swann and a few others in remote viewing the planet Mars. In 2011, after a 25-year hiatus, Tom performed a successful CRV session with Ingo Swann as his monitor, correctly identifying Bridal Veil Falls, New York by name.

A BIOMIND SUPERPOWERS BOOK FROM
SWANN-RYDER PRODUCTIONS, LLC

www.ingoswann.com

Other Books by Ingo Swann

Lightning Source UK Ltd.
Milton Keynes UK
UKHW021440260321
381037UK00008B/2265